The Early Earth

The Early Earth

REVISED EDITION

An Introduction to Biblical Creationism

John C. Whitcomb

BAKER BOOK HOUSE
Grand Rapids, Michigan 49516

ISBN: 0-8010-9679-0

Printed in the United States of America

Original Edition: thirteen printings; 83,866 copies
First printing of the Revised Edition: December 1986
Fifth printing, October 1990

Picture credits
23, 47, 55, 61, 65, 81, 91, 99, 101, 105, 113, 121, 145, 157 — H. Armstrong Roberts
69 — Underwood and Underwood
85 — The American Museum of Natural History
89 — Institut Pasteur Musée
109 — Australian Tourist Commission

To the memory of

my father

Colonel John C. Whitcomb

(1894–1976)

Contents

List of Illustrations

Foreword

To interpret the Bible literally is simply to take God at His Word. This is the high road of biblical exposition which readers have learned to anticipate in anything written by John Whitcomb, and the studies in this book are outstanding examples of such God-honoring exposition.

Any study by Dr. Whitcomb dealing with the Genesis record is certain to be authoritative and relevant. The first edition of this book (thirteen printings) has already been widely read and used to strengthen the confidence of many people in the integrity of God's Word. The early chapters of Genesis should be recognized as completely historical and scientifically accurate, and this is the position demonstrated and expounded in these pages.

I am glad to endorse this book, not only because of the high scholarship and careful exegesis which I know always characterize the writings of Dr. Whitcomb, but also because our many years of close friendship have enabled me to know him as a humble, gracious, and Spirit-filled servant of the Lord Jesus Christ.

I surely hope and believe that this important study, now revised and expanded, will enjoy an even wider ministry than the original edition. It is certainly needed in this day of widespread compromise and retreat and will be commended "by the word of truth, by the power of God" (2 Cor. 6:7).

Henry M. Morris

Preface

Life on the Planet Earth reveals the clear signs of its coming extinction. The drift of quality and order is not upward, but downward. Even as the force of gravity inevitably brings each flying arrow to the ground, so all of nature seems to be programmed to weaken, kill, and disintegrate the fantastically complex and delicately beautiful life forms that have filled the earth in vast abundance.

In spite of the enormous influence of Darwin's theories in modern education, many specialists in the natural sciences are becoming convinced that "nature" could neither have created nor have increased the complexity of the physical universe. The sun dissipates its vast nuclear energy at the stupendous expense of four million tons of its mass per second. And this loss can never be regained! High-level energy systems are inevitably reduced to low-level energy systems, and thus the spectre of a universal "heat-death" darkens the cosmic horizon.

Living systems are similarly trapped in this universal slide toward disorder. Plants, animals, and human beings all lose their original genetic powers through the accumulation of harmful mutations and through the downward gradient of genetic depletion. In opposition to the neo-darwinian concept of inevitable progress, the Bible is in complete harmony with the observed facts of progressive disorder: "The earth . . . and the heavens . . . will perish, but Thou dost endure; And all of them will wear out like a garment; Like clothing Thou wilt change them, and they will be changed. But Thou art the same, And Thy years will not come to an end" (Ps. 102:25–27; cf. Isa. 51:6).

However, strange as it may seem, this very fact of universal deterioration (confirmed and interpreted by God's written revelation) points to man's only true hope of immortality! For if the universe has been evolving into higher and higher forms, as neo-darwinists believe, then the biblical world-view and God's firm promise of eternal salvation to those who believe Him would be hopelessly discredited.

On the other hand, the inexorable grip of the second law of thermodynamics (which states that disorder in a closed system increases with time) forces us to the conclusion that the earth was at one time more organized and integrated and beautiful than it is now. And this in turn points to an infinite and personal God who alone could have infused order and high-level energy into the universe at the beginning.

Christians believe that this great Creator has condescended to tell us what the early earth was really like and how it was brought into being. This record has been preserved for us in that unparalleled account of ultimate origins, the Book of Genesis. What this book tells us about the condition of the earth at the dawn of its existence is of vast importance. For upon this question hinges not only the nature and dignity of man, but also his destiny. The Bible informs us that the perfection of the early earth was an appropriate reflection of man's original fellowship with his Creator. Unfallen man exercised *full dominion* over the earth and its living creatures (Gen. 1:26–28). God put *all things* under his feet and crowned him with glory and honor (Ps. 8:5–8; Heb. 2:5–8). But man rebelled against his gracious Creator and the earth was transformed into its present pattern of frustration, suffering, and death.

> . . . through one man sin entered into the world, and death through sin. . . . For the creation was subjected to futility, not of its own will, but because of Him who subjected it . . . For we know that the whole creation groans and suffers the pain of childbirth together until now (Rom. 5:12; 8:20–22).

But for those among men who repent of sin and turn back to God in obedient faith, the prospect of the future is bright beyond description. After speaking of the removal of the "reign of tooth and claw" that has characterized the animal kingdom since the fall of man, the prophet Isaiah states, "They will not hurt or destroy in all My holy mountain, for the earth will be full of the knowledge of the LORD, as the waters cover the sea" (Isa. 11:9). This revelation concerning the future earth can only be appreciated in the light of biblical revelation concerning the early earth (Gen. 1:29–31).

In identical fashion, the apostle Peter urged the nation of Israel to repent of their rejection of Christ, ". . . in order that times of refreshing may come from the presence of the Lord [i.e., the Kingdom age] . . . the period of restoration of all things about which God spoke by the mouth of His holy prophets from ancient time" (Acts 3:19-21). Since "restoration" (Gr. *apokatastasis*) here means "restoration to a former state," it is obvious that the original perfections of the earth will once again be experienced by those who take God at His Word.

As we see the present earth deteriorating before our very eyes, may this brief study of the early earth arouse within the heart of each reader a deep desire to discover and experience God's announced remedy for the impending crisis: genuine faith in the Lord Jesus Christ as the only Savior of mankind.

Acknowledgments

I hereby express my deep appreciation to Henry M. Morris, for many years the Head of the Department of Civil Engineering at Virginia Polytechnic Institute and since 1970 the Director of the Institute for Creation Research, Santee, California. For over thirty years he has generously shared with me his rich experience in the Word of God and in the natural sciences. Though separated by half a continent in our professional labors, it has been a close association in the joyous task of exploring God's universe in the light of His written revelation. Dr. Morris has read the manuscript in its original form and has written the foreword.

Dr. John J. Davis, Professor of Old Testament and Hebrew, and President of Grace Theological Seminary and Grace College, has provided helpful assistance at various stages of the preparation of the book.

Dr. Donald B. DeYoung, Professor of Physics and Astronomy, Grace College, and Mr. Paul M. Steidl, who have made outstanding contributions to creationist astronomy, have graciously assisted me in writing portions of Chapter 2, "The Creation of the Universe." Dr. DeYoung also assisted in proofreading Chapters 1-2.

Mr. Robert D. Ibach, Jr., Library Director, Dallas Theological Seminary, has read the manuscript in both editions, offering numerous helpful suggestions. His editorial skills are much appreciated.

Dr. Trevor P. Craigen, Coordinator of Biblical Studies, Grace Seminary Extension in Europe, carefully checked all the final page proofs.

Dr. David C. Whitcomb offered helpful suggestions concerning the final page proofs.

Mrs. LeAnne Christianson generously gave of her time and skill in preparing the typescript for publication.

Mr. Dan Van't Kerkhoff and Mr. Gordon DeYoung have graciously assisted me in the various stages of this publishing project.

My dear wife, Norma, has made a major contribution to this book through her prayers and encouragement.

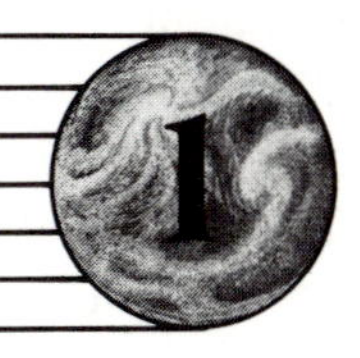

The Nature of Biblical Creation

Creation Was Supernatural

In forthright opposition to all efforts to explain the origin of the world in terms of purely natural processes, the Bible states that *God* created all things *supernaturally*. In other words, the world came into being in a way that was entirely different from anything that may be observed in the present universe. Today absolutely nothing is being created directly apart from preexistent materials, and scientists express this basic truth in terms of the first law of thermodynamics (i.e., energy can be neither created nor destroyed). Genuine creation is no longer being accomplished, as the Bible clearly states (Gen. 2:1–3). God's work of *preservation* keeps the universe in existence (Heb. 1:3), and His work of *providence* directs the universe toward glorious goals (Col. 1:20), but His work of *creation*, as far as the present universe is concerned, has been completed.

Thus, when God created "the heavens and the earth, the sea and all that is in them" (Exod. 20:11; cf. 31:17; Neh. 9:6), He did so without the use of any preexistent materials whatsoever. In one moment there was no physical substance anywhere; in the next moment the heavens and the earth sprang into existence. Theologians have called this *creatio ex nihilo* (creation out of nothing), and this expression is helpful if we understand it to mean that *physical* entities were created out of the nonphysical resources of God's omnipotence. Technically, the expression is applicable only to the creation of inorganic substances, for God did employ previously created inorganic materials in forming the bodies of living things. Nevertheless, even in this case, as we shall see, creation was strictly supernatural.

The fact that creation was supernatural means, among other things, that it can be grasped by the human mind only through the channel of *special revelation*. God alone can tell us how the world began, because no man was there to see it being created, and even if a human observer had been present, he could not have understood fully what he saw apart from God's own interpretation. "Now gird up your loins like a man," said God to Job, "And I will ask you, and you instruct Me! Where were you when I laid the foundation of the earth! Tell Me, if you have understanding" (Job 38:3–4).

However, our difficulty in grasping the doctrine of creation is not due so much to the fact that we are *finite* as to the fact that we are *sinful*. "But a natural man does not accept the things of the Spirit of God; for they are foolishness to him, and he cannot understand them, because they are spiritually appraised" (1 Cor. 2:14). There are few doctrines of the Bible that seem more foolish to the natural man than that of supernatural creation, for such events are not happening today. But creation is most definitely one of the supremely important "things of the Spirit of God," for without it the Scriptures and Christianity would fall to pieces. Remove this doctrine, and the entire superstructure collapses.

It is therefore exceedingly important that we approach the first chapters of Genesis in the light that God Himself provides through the entire testimony of Scripture. Even as God commanded Moses to put off his shoes because the place whereon he stood was holy ground, so likewise we must set aside our concepts of what could or could not have happened, and stand in God's presence, ready to hear and to believe what He has chosen to tell us about creation.

Such unconditional submission to the authority of the Word of God is not, of course, the mood of our day, even among Christians. Paul warned that

> . . . the time will come when they will not endure sound doctrine; but wanting to have their ears tickled, they will accumulate for themselves teachers in accordance to their own desires; and will turn away their ears from the truth, and will turn aside to myths (2 Tim. 4:3–4).

One such "myth," we believe, is that God did not create the world supernaturally, but employed natural processes, by His providence, through vast periods of time. It is a myth, not simply because it contradicts Scripture, but because it contradicts the observable processes of the universe.

In recent years remarkable testimony has come from the pens of highly respected scientists who recognize that the evolution concept, in its broader aspect, rests upon a vanishing foundation. G. A. Kerkut of the Department of Physiology and Biochemistry at the University of Southampton, for example, notes that evolutionists often write as though they "have had their views by some sort of revelation." In spite of "many gaps and failures" in their system, it is "taken on trust" by a "blind acceptance" and a "closing of the eyes" to many important facts, thus revealing an "arrogant" rather than truly scientific spirit.[1] Attempts to bridge the gap between invertebrates and vertebrates for example, have resulted in "science fiction" rather than discovery,[2] and the possibility that life first began spontaneously is a "matter of faith on the part of the biologist."[3]

In his introduction to Darwin's *The Origin of Species* in the *Everyman Library* (1956), W. R. Thompson points out that

> Modern Darwinian palaeontologists are obliged, just like their predecessors and like Darwin, to water down the facts with subsidiary hypotheses, which, however plausible, are in the nature of things unverifiable . . . and the reader is left with the feeling that if the data do not support the theory they really ought to. . . . This situation, where scientific men rally to the defence of a doctrine they are unable to define scientifically, much less demonstrate with scientific rigour, attempting to maintain its credit with the public by the suppression of criticism and the elimination of difficulties, is abnormal and undesirable in science.[4]

Several years ago, Richard B. Goldschmidt (1878–1958), a leading geneticist, stated,

> The incessant repetition of this unproved claim (of micromutational evolution), glossing lightly over the difficulties, and the assumption of an arrogant attitude toward those who are not easily swayed by fashions of science, are considered to afford scientific proof of the doctrine.[5]

---

1. *Implications of Evolution* (New York: Pergamon Press, 1960), pp. 154, 155.
2. Ibid., p. 153.
3. Ibid., p. 150.
4. Reprinted in *Journal of the American Scientific Affiliation* 12:1 (March 1960), pp. 7, 8.
5. "Evolution, As Viewed by One Geneticist," *American Scientist*, January 1952, p. 94.

The Planet Earth

The early earth, as it would be viewed from outer space before the great Flood, was quite different from its present appearance. In the first place, it would have been even more colorful than it is now, for there would have been no cloud canopy to obscure the brillant blue oceans. Secondly, there would have been no white polar caps or reddish-brown desert regions, for thick green vegetation covered almost all of the land areas, even in polar regions (thick coal deposits have been discovered in the mountains of Antarctica). And thirdly, the continents were probably quite different in shape and location. Some regions that are now high above sea level were once under the oceans.

Many Bible scholars believe that there was only one great land mass surrounded by seas before the Flood, because two each of all kinds of air-breathing animals walked to Noah's ark (Gen. 6:20; 7:8). It is also possible, however, that if there had been more than one continent, representatives of all the kinds of animals would have been living on the continent where the Ark was being constructed. The idea that continents have "drifted" into their present locations faces serious geophysical objections and is not really supported by Scripture (Gen. 10:25 must refer to the dividing of nations after the Tower of Babel judgment; cf. 10:5, 20, 32).

J. J. Duyvené de Wit of the Department of Zoology at the University of the Orange Free State, pointed out that the "dualistic split" between scientific *knowledge* (pertaining to discontinuity between kinds of living things) and suprascientific *faith* (in evolutionary continuity) amounts to "a rift in the consciousness of the biologist personally."[6]

The general theory of evolution, therefore, as an anti-theistic faith, has been increasingly contradicted by the realities of empirical science during the past century. Christians who accept the clear testimony of Scripture concerning the supernatural character of original creation are confident that the true facts of science, though frequently suppressed and misinterpreted by evolutionists, will ultimately be found to harmonize with the Bible.[7]

Creation Was Sudden

The creation of the astronomical universe was not only *ex nihilo* (i.e., from no previously existing matter, as stated in Hebrews 11:3[8]), but was also, for this very reason, *instantaneous*. The origin of universal mass/energy and the various force fields (such as gravitation) could not, therefore, have been spontaneous or self-acting. The evolutionary concept of a gradual build-up of heavier and heavier elements throughout billions of years is clearly excluded by the pronouncements of Scripture.

In the first place, the immediate effect of God's creative word is emphatically stated by the psalmist:

> By the word of the LORD the heavens were made, and by the breath of His mouth all their host. . . . Let all the earth fear the LORD; let all the inhabitants of the world stand in awe of Him. For He spoke, and it was done; He commanded and it stood fast (Ps. 33:6–9; cf. Ps. 148:1-6).

6. A *New Critique of the Transformist Principle in Evolutionary Biology* (Kampen, Neth.: Kok, 1965), p. 43. "Throughout the past century there has always existed a significant minority of first-rate biologists who have never been able to bring themselves to accept the validity of Darwinian claims. In fact, the number of biologists who have expressed some degree of disillusionment is practically endless" (Michael Denton, *Evolution: A Theory in Crisis* [Bethesda, MD: Adler & Adler, 1986], p. 327).

7. Cf. Henry M. Morris, *Creation and the Modern Christian* (El Cajon, CA: Master Book Publishers, 1985); and *History of Modern Creationism* (El Cajon, CA: Master Book Publishers, 1984).

8. Hebrews 11:3 certainly cannot mean that the physical substances which compose our visible universe consist of "invisible" atomic particles. Spiritual faith is not required to accept the atomic theory of matter. The point of this key statement on creationism is that visible, material substances did not exist in any form whatsoever, other than in the mind of an eternal and omniscient God, until He spoke His creative word.

There is certainly no thought here of gradual development, or age-long, step-by-step fulfillment of God's command. In fact, it is quite impossible to imagine any time interval in the transition from absolute nonexistence to existence!

Similarly, ". . . God said, 'Let there be light'; and there was light" (Gen. 1:3). At one moment, there was no light anywhere in the universe. At the next moment, light existed! So spectacular is this particular creation event that the New Testament compares it to the *suddenness* and *supernaturalness* of spiritual conversion: "For God, who said, 'Light shall shine out of darkness,' is the One who has shone in our hearts to give the light of the knowledge of the glory of God in the face of Christ" (2 Cor. 4:6; cf. 5:17). Also, God is able to raise *suddenly* the physical dead, because He is the God who "calls into being that which does not exist" (Rom. 4:17). It may be confidently asserted that the idea of *sudden appearance* dominates the entire creation account (cf. Gen. 1:3, 12, 16, 21, 25, 27; 2:7, 19, 22).[9]

There are many today who deny this important biblical concept out of respect for the supposed requirements of empirical science. But there is absolutely nothing in empirical science that prevents the living God, who sustains the observable and measurable processes of "empirical science" in His hand moment by moment, from changing His methods from time to time to accomplish His eternal purposes for men. From a biblical perspective, as will be demonstrated in the pages that follow, the evidence is overwhelming that God's creative and redemptive programs are characterized by sudden and supernatural initiatory events.

At the same time, it must be emphasized that the startling suddenness of God's *supernatural* creative acts and sign-miracles is never intended in Scripture to minimize the glory of God's non-miraculous *providential* works in human history (cf. Dan. 4:17 and the Book of Esther[10]). The difference between these two manifestations of God's sovereign control of His world is highly significant. Miracle and providence are *not* identical and dare not be confused. The conception of our Lord Jesus Christ, for example, was both *sudden* and *supernatural* while His birth was

9. Russell W. Maatman, a proponent of the Day-age Theory, was impressed with this biblical evidence. "There is no doubt that *each creation event* was instantaneous. One moment a certain thing existed; the previous moment, it did not exist" (*The Bible, Natural Science and Evolution* [Grand Rapids: Baker Book House, 1970], p. 95.

10. Cf. John C. Whitcomb, *Esther: The Triumph of God's Sovereignty* (Chicago: Moody Press, 1979).

the result of a *gradual* and *natural* process carried out under the providential control of God. If the conception of Christ is understood to be providential rather than miraculous, the incarnation is denied and Christianity is destroyed (cf. 1 John 4:3; 2 John 7). Likewise, if the events of Genesis 1–2 are understood to be providential rather than miraculous, biblical creationism is not simply modified; it is destroyed.

Creation *ex nihilo* refers primarily to angels (cf. Col. 1:16), the astronomic universe (with all of its complexities of visible objects and invisible force fields), and this planet. When God created living things on the earth, however, He formed them suddenly from previously created inorganic substances. Thus, He commanded the waters to bring forth marine and flying creatures on the fifth day. However, the water, by itself, even in the presence of sunshine, could *never* (even in billions of years!) have brought forth such marvelously complex and beautiful animals. By the same token, the water used by our Lord at Cana of Galilee (cf. John 2:1–11) could never have turned into wine, even if it vibrated with evolutionary anticipation in those stone jars for billions of years. In both cases, complex entities appeared suddenly, even though built upon preexistent lifeless materials. Thus, the fact that God commanded the earth to bring forth trees no more implies a gradual growth process than His use of the same inorganic elements to bring forth the full-grown body of a man at the end of creation week. Even with regard to the origin of the human race, many Christians have seen divine providence through time and process instead of divine miracle, and thus have twisted the Genesis record out of recognition. This will be discussed further in chapter 4.

One writer, a theistic evolutionist who rejects all creation miracles, has characterized the "suddenness" interpretation of Genesis 1–2 as being dangerously close to the theology of the pagan Ephesians who believed that the image of Diana had hurtled upon them out of heaven![11] Typical of this type of creationism, we are told, is the modern fundamentalist movement which "is ill at ease in the presence of the processive" and thus assigns a large place "to the idea of the Second Coming, looked upon as in no sense the fruition of the historic process, but as something brought about simply and solely by the interrupting voice of God."[12]

The validity of this type of objection rests, of course, upon the validity

11. Leonard Verduin, "Man, A Created Being: What of an Animal Ancestry?" *Christianity Today* (May 21, 1965), p. 10.

12. Ibid., p. 11.

of the assumption that neo-darwinian evolutionism is true, that biblical miracles can usually be explained in terms of providential processes, and that God created the world "with the prodigal disregard for the passing of time that marks the hand of him who fashions a work of art."[13]

This leads us to a second major consideration pertaining to the suddenness of creation events in Genesis, namely, the analogy of God's creative works in the person of His Son during His earthly ministry nearly two thousand years ago in Palestine. The New Testament clearly teaches us that the entire universe was created through God's Son (John 1:3, 10; Col. 1:16; Heb. 1:2). The New Testament further reveals that the works He performed during His brief earthly ministry were intended to reveal His true nature and glory (John 1:14; 2:11; 20:31). In the light of these truths, it is profoundly instructive to observe that *all* of Christ's miracles involved *sudden transformations*.

Though someone has stated that there is "no strategy as slippery and dangerous as analogy," the biblical analogy of Christ's creative work in Genesis and in the Gospels remains overwhelmingly and decisively powerful. In response to the mere word of our Lord, for example, a raging storm *suddenly* ceased, a large supply of loaves and fishes *suddenly* came into existence, a man born blind *suddenly* had his sight restored, and a dead man *suddenly* stood at the entrance of his tomb. Of the vast number of healing miracles performed by Christ, the only recorded exception to instantaneous cures is that of the blind man whose sight was restored in two stages, each stage, however, being instantaneous (Mark 8:22–25).[14]

Such miracles were undeniable signs of supernaturalism in our Lord's public claim to be Israel's Messiah, and we may be quite sure that if, in His healing of the sick, crippled, and blind, He had exhibited "the prodigal disregard for the passing of time that marks the hand of him who fashions a work of art,"[15] no one would have paid any attention to His claims. If the Sea of Galilee had required two days to calm down after

13. Ibid., p. 10.

14. This remarkable exception to our Lord's many thousands of instantaneous cures certainly cannot be used as a basis for progressive creationism, with its step-by-step (and yet supernatural) concept of origins. The very fact that this case is singled out in Scripture actually serves as a warning to those who would assume that this was God's basic method of creating things! The Genesis account gives absolutely no hint of multi-stage "creations" of angels, stars, planets, plants, animals, or human beings throughout millions or billions of years.

15. Ibid.

Jesus said, "Peace, be still!" the disciples would neither have "feared exceedingly," nor would they have "said one to another, 'What manner of man is this, that even the wind and the sea obey Him!'" (Mark 4:39–41, KJV).

The enormous theological significance of these facts for a proper Christian understanding of the origin of the world can be recognized in the following observation:

> The theologian attributes certain *infinite* properties to his God; he is described as omnipotent, omniscient, and of infinite goodness. Now the Mind which reveals itself in the development of life on this planet is clearly not omnipotent; otherwise it would have assembled perfectly designed organisms directly from the dust of the earth without having to go through the long process of trial and error which we call evolution.[16]

Every effort to modify the suddenness and supernaturalness of creation events in order to make them more acceptable to the "modern mind" only results, in the long run, in minimizing and obscuring the true attributes of the God of creation. This has been a difficult lesson for many Christians to learn.

The third major consideration is the fact that God's work of creation was completed in six literal days, thus demonstrating conclusively that His creative work during each of these days was both sudden and supernatural. In view of the widespread resistance to this concept, even in some Christian circles, it may be surprising to many people to learn how strong are the biblical evidences in its support, if the indispensable and time-honored historical/grammatical system of biblical interpretation be accepted.

Four evidences for a literal seven-day creation week will now be presented, along with answers to major objections.

1. Although the Hebrew word for "day" (*yôm*) is used nearly two thousand times in the Old Testament, only in rare cases can it refer to a time period longer than twenty-four hours, and then only if the context demands it (e.g., "day of the Lord"). However, when a *numerical adjective* is attached to the word "day" (two hundred known cases in the OT) its meaning is *always* restricted to twenty-four hours (i.e., "first day," "second day," etc., with a precise parallel in Numbers 7:12–78). Over seven hun-

16. John L. Randall, *Parapsychology and the Nature of Life* (London: Souvenir Press, 1975), p. 235.

dred times the plural form "days" (*yāmim*) appears in the Old Testament, and it *always* refers to literal days (e.g., Exod. 20:11—"in six days").[17] The expression "one day" in Zechariah 14:7 (KJV), claimed by some to be an exception to this rule, must refer to a literal day also, especially because the term "evening" appears in the same verse.

Four times within the creation narrative the word "day" refers to the twelve-hour period of daylight (1:5, 14, 16, 18), but no numerical adjectives are used and the context clearly shows which sense is being used (which is true in English as well). For example, the terms "day" and "night" in Genesis 1:5 are described as periods of "light" and "darkness." This would be utterly meaningless here if "day" and "night" are not parts of a normal day. In Genesis 1:14–19, the sun was created to "govern the day" and the moon to "govern the night." Again, "day" and "night" here must refer to parts of a normal day. The expression "in the day" (*bᵉ yôm*) in Genesis 2:4 not only lacks the numerical adjective but becomes, by its attached preposition "bᵉ," an idiom for "when."[18]

Robert C. Newman and Herman J. Eckelmann, Jr., who reject the literal-day interpretation, nevertheless concede that "no clear counter-example [of *yôm* with an ordinal number] can be cited with *yôm* meaning a long period of time."[19] They further make the very damaging concession that "the most common meanings of the words involved [e.g., "day," "evening," "morning"] should be used in constructing a model."[20]

As a last resort, it seems, proponents of the Day-age Theory can only say that "the absence of the use of 'yamim' for other than regular days and the use of ordinals only before regular days elsewhere in the Old

17. Cf. Robert E. Kofahl and Kelly L. Segraves, *The Creation Explanation* (Wheaton, IL: Harold Shaw Publishers, 1975), pp. 321–32. Note also the listing of every appearance of the word "day" in the Old Testament in Robert L. Thomas, ed., *New American Standard Exhaustive Concordance of the Bible* (Nashville: Holman, 1981), pp. 277ff.

18. Cf. Francis Brown, S. R. Driver and Charles A. Briggs, eds., *A Hebrew and English Lexicon of the Old Testament* (Oxford: Clarendon Press, 1975), p. 400. The term *bᵉ yôm* with an infinitive construct lacks the chronological sharpness of *yôm* with a numerical adjective. A total of fifty-six occurrences of the *bᵉ yôm* usage is very strong attestation for the translation "when" or "at the time when" in Genesis 2:4. Cf. Allen P. Ross, "Genesis," in J. F. Walvoord and R. B. Zuck, eds., *The Bible Knowledge Commentary: Old Testament* (Wheaton, IL: Victor Books, 1985), p. 30.

19. Robert C. Newman and Herman J. Eckelmann, Jr., *Genesis One and the Origin of the Earth* (Downers Grove, IL: InterVarsity Press, 1977), p. 61.

20. Ibid., p. 74.

Testament cannot be given an unequivocal exegetical significance in view of the uniqueness of the events being described in Genesis 1 (i.e., sequential, indefinite time periods)."[21] Perhaps their argument can be restated as follows: God's revelation to us through science indicates very clearly that plants and animals have been living and dying for millions and billions of years; therefore, exegetical support for the traditional Hebrew/Christian understanding of a literal and relatively recent creation week cannot be determinative. One wonders how many other philosophically "unacceptable" portions of Scripture can be dismissed in this way.

2. The qualifying phrase, "the evening and the morning," which is attached to each of the creation days throughout Genesis 1, indicates a twenty-four-hour cycle of the earth rotating on its axis in reference to a fixed astronomical light source (not necessarily the sun in every case). The same phrase appears in Daniel 8:26 (cf. 8:14 NASB, NIV), where it simply cannot mean long, indefinite periods of time. Some have claimed Psalm 90:6 as an example of a figurative use of "evening" and "morning." However, the Genesis 1 formula is not used here, and the order of the words is reversed. Furthermore, the figurative use of these terms in Psalm 90 would be meaningless if they did not presuppose a literal use in earlier historical narratives of Scripture, such as Genesis 1.

3. A creation "week" of six indefinite periods of time would hardly serve as a valid meaningful pattern for Israel's cycle of work and rest, as explained by God at Sinai in the fourth commandment (Exod. 20:11; 31:17).

While it is, of course, true that God *could* have created the world in six billion years, or in six seconds (or in a split second!) if He had chosen to do so, such speculations are completely irrelevant in the face of the fourth commandment which informs us that God, as a matter of fact, chose to create the world "in six days" in order to provide *a clear pattern* for Israel's work periods and rest periods. The phrase "six days" (note the plural) can hardly be figurative in such a context.

Leon Morris also presents an interesting parallel between the creation week and the first week of Christ's public ministry:

21. Walter L. Bradley and Roger Olson, "The Trustworthiness of Scripture in Areas Relating to Natural Science," in Earl D. Radmacher and Robert D. Preus, eds., *Hermeneutics, Inerrancy, and the Bible* (Grand Rapids: Zondervan Publishing House, 1984) p. 299. In the same volume, Gleason L. Archer goes so far as to state that "the twenty-four hour theory was never correct and should never have been believed" (p. 329). See the response by Henry M. Morris in the same volume, pp. 337–48.

> If we are correct in thus seeing the happenings of one momentous week set forth at the beginning of this Gospel we must go on to ask what significance is to be attached to this beginning. The parallel with the days of creation in Genesis 1 suggests itself, and is reinforced by the "In the beginning" which opens both chapters. Just as the opening words of this chapter recall Genesis 1, so it is with the framework. Jesus is to engage in a new creation. The framework unobtrusively suggests creative activity.[22]

4. Since the word "days" in Genesis 1:14 is linked with the word "years," it is quite obvious that our well-known units of time are being referred to, their duration being determined not by cultural or subjective circumstances, but by the fixed movements of the earth in reference to the sun. Otherwise the term "years" would be meaningless.

We must assume that the first three days of creation week were the same length as the last three astronomically fixed days, because exactly the same descriptive phrases are used for each of the six days (i.e., numerical adjectives and the evening/morning formula), and all six days are grouped together in Exodus 20:11 to serve as a model for Israel's work week. The fact that the sun was not created until the fourth day does not make the first three days indefinite periods of time, for on the first day God created a fixed and localized light source in the heaven in reference to which the rotating earth passed through the same kind of day/night cycles as it has since the creation of the sun.[23]

This light could not have been the light of God's divine nature, for it was created by God ("let there be"). Furthermore, if it had been the light of God Himself half of the earth would not have remained in darkness. Thus, it was a created light, localized somewhere in the universe, and possibly incorporated into the stellar universe on the fourth day of creation week after its unique and temporary function was accomplished. The theological implications of the postponement of the creation of the sun, moon, and stars to the second half of creation week are highly significant in view of the idolatrous use of these finite, inanimate objects by fallen man (cf. Deut. 17:3; Job 31:26–28).[24]

Thus, a careful analysis of the Hebrew term "day" in its Old Testament

22. *The Gospel According to John* (Grand Rapids: Wm. B. Eerdmans Publishing Company, 1971), p. 130.

23. Cf. H. C. Leupold, *Exposition of Genesis* (Columbus, OH: The Wartburg Press, 1942), pp. 51–53.

24. See John C. Whitcomb and Donald B. DeYoung, *The Moon: Its Creation, Form and Significance* (Winona Lake, IN: BMH Books, 1978), pp. 153–62.

usage, in conjunction with the numerical adjective and the evening/morning formula, and in conjunction with the term "years," especially in the light of Israel's God-given cycle of work and rest, leads us inevitably to the conclusion that the days of creation were literal and successive 24-hour days. To stretch the days into long ages or to insert long ages between the days is biblically illegitimate.[25] The traditional Judeo-Christian understanding is thus confirmed by biblical exegesis: *the universe was created by God within one literal week of days.*

In opposition to the literal-day interpretation of Genesis 1, it has been argued that other passages of the Bible speak of a "day" in God's sight being as a thousand years. It is true that one such statement is found in the Old Testament (Ps. 90:4) and one in the New Testament (2 Peter 3:8). But so far from weakening the literal-day view, these verses actually help to strengthen it.

In 2 Peter 3:8, for example, we are *not* told that God's days last a thousand years each, but that "with the Lord one day is as a thousand years. . . ." To say "*as* a thousand years" is a very different matter from saying "*is* a thousand years." This point has often been overlooked. If "one day" in this verse really means a long period of time, then we would end up with the following absurdity: "with the Lord a long period of time is as a thousand years." But a thousand years would be a long period of time for human beings too! Psalm 90:4 must be understood in the same way if the proper contrast between God and man is maintained: "For a thousand years in Thy sight are like yesterday when it passes by. . . ." Here "yesterday" *must* refer to a 24-hour period or the contrast is destroyed.

The obvious teaching of Psalm 90:4 and 2 Peter 3:8, then, is that God is above the limitations of time. One valid deduction from this fact is that *God can accomplish in one brief, literal day what man could not accomplish in a thousand years, if ever.* This is one of the astounding messages that comes through to us from a normal interpretation of the creation narrative of Genesis 1: *God alone has infinite power*! The prophet Jeremiah saw this profound truth: "Ah Lord God! Behold, Thou hast made the heavens and

25. For a discussion of the Gap Theory of Genesis 1:1–2 and the Chaos/Creation Theory of Genesis 1:1–3, see chapter 5. Every effort to accommodate the long ages of evolutionary geology, whether before, during, or between the days of creation hopelessly compromises the biblical concept of the curse and death coming into the world only after man's rebellion (cf. Rom. 5:12; 8:18–23).

the earth by Thy great power and by Thine outstretched arm! Nothing is too difficult for Thee! (Jer. 32:17).

Did the Creator, perhaps, really need the seventh day to rest from six days of creative work? The answer comes back to us with overwhelming clarity: "Do you not know? Have you not heard? The Everlasting God, the LORD, the Creator of the ends of the earth does not become weary or tired. His understanding is inscrutable. He gives strength to the weary, and to him who lacks might He increases power" (Isa. 40:28–29).

There is, in fact, simply no way for the human mind to grasp the power of God: "'To whom then will you liken Me that I should be his equal?' says the Holy One. . . . 'For as the heavens are higher than the earth, so are My ways higher than your ways, and My thoughts than your thoughts'" (Isa. 40:25; 55:9).

This enormously significant truth concerning God is seriously compromised if not destroyed when the creation account in Genesis is stretched to incorporate vast ages of time in order to make the passage more "reasonable" and "scientifically credible," and thus to accommodate it to man's finite, naturalistic level of thinking. To twist the Scripture is to distort God's message to us. What the apostle Peter said concerning Paul's letters is surely applicable to the opening chapters of the Bible, ". . . in which are some things hard to understand, which the untaught and unstable distort, as they do also the rest of the Scriptures, to their own destruction" (2 Peter 3:16).

Another widely held objection to the literal-day interpretation of Genesis 1 is that the seventh day has not yet terminated, for God is still resting from His work of creation (cf. Heb. 4:3–11).[26]

This argument introduces much confusion between historical events and their spiritual application. The "rest" of Hebrews 4 is primarily the spiritual rest of salvation (cf. Matt. 11:28–30), whereby the believer shares in the eternal blessedness and fulfillment which characterizes God. Certainly God did not have to wait until the end of the sixth day of creation week for this kind of rest to begin! Thus, the first sabbath was not instituted for God's benefit (cf. Isa. 40:28; John 5:17) but rather for man's benefit (Mark 2:27). *It is this often neglected point that is crucial in determining the duration and purpose of the original sabbath day.*

26. Cf. R. C. Newman and H. J. Eckelmann, *Genesis One and the Origin of the Earth*, pp. 65, 85.

Edward J. Young insisted that "the seventh day is to be interpreted as similar in nature to the preceding six days. There is no Scriptural warrant whatever (certainly not in Hebrews 4:3–5) for the idea that the seventh day is eternal."[27] Homer A. Kent, Jr., concurs:

> This [the fact that no terminus is mentioned for the seventh day] does not imply that the seventh day was not a literal day with an evening and a morning, just as the previous six days of creation. However, the author has used the silence of Scripture on this point to illustrate his argument that God's sabbath rest has never ended. The same method or argument is used in [Hebrews] 7:3 regarding Melchizedek's absence of recorded birth, parentage, or death.[28]

How long, then, did the first sabbath continue? It is clear that all Israelites, to whom sabbath observance was specifically assigned by God, understood this period to be exactly twenty-four hours in length, based on the pattern of God's creation sabbath:

> Six days you shall labor and do all your work, but the seventh day is a sabbath of the LORD your God; in it you shall not do any work. . . . For in six days the LORD made the heavens and the earth, the sea and all that is in them, and rested on the seventh day; therefore the LORD blessed the sabbath day and made it holy (Exod. 20:9–11).

Any Israelite who decided to extend his sabbath observance indefinitely on the assumption that God's sabbath still continued would have starved to death (cf. Exod. 35:3). Equally significant is the deduction that Adam and Eve must have lived through the entire seventh day of creation week before God expelled them from the garden, for God would not have cursed the ground (Gen. 3:17) during the very day He "blessed" and "sanctified" (Gen. 2:3).

Some have even argued that the sixth day of creation must have lasted much longer than twenty-four hours because God must have given Adam

27. Edward J. Young, *Studies in Genesis One* (Phillipsburg, NJ: Presbyterian and Reformed Publishing Company, 1964), p. 77, note 73.

28. Homer A. Kent, Jr., *The Epistle to the Hebrews* (Winona Lake, IN: BMH Books, 1972), p. 82, note 32. Charles C. Ryrie agrees: "Strictly speaking, Hebrews does not say other than [that] God rested on the seventh day. It does say He rest*ed*, not He rests" (*Basic Theology* [Wheaton, IL: Victory Books, 1986], p. 186).

enough time to become lonely! This is supposedly confirmed by the fact that when he was awakened and beheld Eve, he exclaimed: ". . .This at last [*happa 'am*] is bone of my bones and flesh of my flesh" (Gen. 2:23a, RSV)[29].

But surely the term *happa 'am* ("this once" or "now at length")[30] cannot be pressed to mean an absolutely (instead of a relatively) long period of time. Jacob could appropriately have used this expression after two or three hours of intense wrestling with God (Gen. 32:24; Hos. 12:2–5). But it is not necessary to speculate about the possible uses of the term, for in Genesis 18:32 Abraham used it at the end of a single conversation with his Lord! Newman and Eckelmann explain: "here the strong emotional climax may build quickly because Abraham is bargaining with God."[31]

This explanation clearly negates their entire argument, for Adam would hardly have been less emotionally involved with his God and his newly-created wife than Abraham was with his God. Sadly, Christian thinkers who insist that Adam could *not* have named the birds and mammals in one day, even with a freshly-created and unfallen mind, and with God's special help (Gen. 2:19—"the LORD God . . . brought them to the man"), are indulging in the typically hazardous uniformitarian extrapolations that characterize so many studies of Genesis 1–11 in our generation. Giving a little more time to Adam to name the birds and mammals may seem to some to be a very inconsequential issue. What Newman and Eckelmann are really asking for, however, is not a few additional days or weeks for Adam to name the animals, but a "creation week" lasting "15 billion to 20 billion years," including evolutionary processes and over a billion years of death in the animal kingdom before the Fall, with the seventh day yet future![32]

In a biblical doctrine as vitally important as the creation of the world, it would seem incredible that God would wait until the nineteenth century A.D. (more than three millenniums after Genesis was written!) to reveal to His people that the creation account actually encompasses vast ages of death and destruction in the animal kingdom before Adam's creation. As one proponent of the ancient-earth concept boldly states:

29. Robert C. Newman and Herman J. Eckelmann, *Genesis One*, p. 131.

30. Cf. Brown, Driver, and Briggs, *Hebrew and English Lexicon of the Old Testament*, p. 822, 2d.

31. Newman and Eckelmann, *Genesis One*, p. 133.

32. Ibid., pp. 83–85.

> Until the end of the eighteenth century, Christians were virtually unanimous in the belief that the Earth was about six thousand years old according to the teaching of Scripture. However, increased scientific study . . . brought pressure to bear upon Christian thinkers to reevaluate the question of the age of the Earth.[33]

To be sure, a small number of church fathers were sufficiently influenced by the pagan philosophies of their day to treat much of Genesis 1 allegorically, just as many theologians do in our day. But, "They frequently toyed not with the idea of creation extending over a long period of time but rather with the idea that it was instantaneous. This position was suggested by Origen, Hillary, Augustine, and Jerome."[34] However, the true teaching of Genesis 1, stripped of all allegorical speculations, comes through to God's people with amazing clarity. Even theological liberals, for their own reasons, emphasize this point. For example, James Barr of the Oriental Institute of the University of Oxford, comments:

> So far as I know, there is no professor of Hebrew or Old Testament at any world-class university who does not believe that the writer(s) of Genesis 1–11 intended to convey to their readers the idea that creation took place in a series of six days which were the same as the days of 24 hours we now experience.[35]

A number of references in the New Testament strongly imply that the human race was created at approximately the same time as the material universe. For example, our Lord stated that "from the beginning of creation, God MADE THEM MALE AND FEMALE" (Mark 10:6). But if billions of years elapsed between the creation of the earth and the creation of the human race, this would be a misleading statement. For similar statements, see Matthew 13:35, Mark 13:19, Luke 11:50, Romans 1:20, Hebrews 4:3, 9:26.[36]

33. Davis A. Young, *Christianity and the Age of the Earth* (Grand Rapids: The Zondervan Corporation, 1982), p. 13.

34. John W. Klotz, *Studies in Creation* (St. Louis: Concordia Publishing House, 1985), p. 68.

35. Personal letter to David C. C. Watson, dated April 23, 1984.

36. Suggested in a personal communication from David C. C. Watson dated April 23, 1984.

The writings of Davis A. Young illustrate the tremendous tensions that are created when uniformitarian (rather than catastrophist) views of historical geology are given equal authority with the early chapters of Genesis. On the one hand, Davis Young insists that Scripture is far more authoritative than science,[37] and that

> . . . if it can be demonstrated beyond all doubt that Scripture demands a 24-hour view of the days, then the Christian scientist must accept that and, in effect, give up geological science and turn to something else. If he is consistent in his faith in Scripture, he must do this.[38]

On the other hand, he continues, ". . . as a geologist, I am quite delighted with this interpretation [the day-age views of Genesis 1], for I have become accustomed to thinking in terms of billions of years!"[39] The figurative view of Genesis 1 "gives the scientist great freedom,"[40] and leaves him "unfettered."[41] "In short, the Christian ought to be willing to let science advance in its own way and in its own time, that is, to develop naturally as new discoveries are made. One cannot force scientific thinking to advance along a particular line."[42] How this approach honors the supremacy of Scripture or differs from straightforward theistic evolutionism is not convincingly explained.

There is simply no escaping the fact that God intends us to understand the creation of the universe and of the earth to have been, for all practical purposes, instantaneous. The implications of this aspect of divine revelation with regard to currently popular attempts to harmonize Genesis with cosmic evolutionism should be perfectly obvious. To suggest a "gradual creation" of the sun, moon, and earth may be conceivable to some minds. But for most people, such a concept would raise the very serious question as to whether God, as a matter of fact, ever really *created*

37. Davis A. Young, *Creation and the Flood: An Alternative to Flood Geology and Theistic Evolutionism* (Grand Rapids: Baker Book House, 1977), pp. 18–22.

38. Ibid., p. 82.

39. Ibid., p. 91.

40. Ibid., p. 87.

41. Ibid., p. 113.

42. Ibid., p. 114. For further analysis of Davis Young's views, see John C. Whitcomb, "The Science of Historical Geology," *The Westminster Theological Journal* 36:1 (Fall, 1973), pp. 65–77; and Donald B. DeYoung, review of Davis A. Young, *Christianity and the Age of the Earth*," in *Grace Theological Journal* 4:2 (Fall, 1983), pp. 297–301.

The Creation Week	
First Day The Son of God (John 1:3, Col. 1:16) created the entire universe. In the *third* heaven (cf. 2 Cor. 12:2) all angelic beings were created righteous and rejoiced to see earth's creation moments later (Job 38:4–7). The *second* heaven of outer space was totally empty and therefore dark, until a temporary localized astronomic light was created to begin the day/night cycle. The *first* heaven of atmospheric gases was apparently cloudless (cf. Gen. 1:6). At this stage the earth was perfect but incomplete ("God created . . . waste and void").	**Second Day** Part of the earth's shoreless ocean was lifted above the atmospheric "firmament" (expanse) to form the canopy that trapped reflected long-wave solar heat radiation. This pre-Flood "greenhouse effect" best explains the clear evidence of vast animal and plant life (much of it tropical) which now lies frozen and buried in arctic zones. The firmament (like the heavens, earth, darkness, and mankind) was not specifically called "good." But all these things were included in the "very good" of 1:31.
Fifth Day Marine and flying animals were created out of the waters in abundance. Whales (largest of the "sea-monsters"—*tannînīm* did not evolve from land mammals, nor did birds evolve from reptiles, for they were created one day earlier. This reversal of order (cf. earth before sun and fruit trees before marine creatures) destroys the day-age view which stretches the days of creation to accommodate the vast time periods of evolutionary geology.	**Sixth Day** Creation of all land animals including dinosaur kinds (among the "creeping things"—cf. Job 40:15–23). Adam was created as an adult and was commissioned to a life of delightful work and willing obedience. He named all bird and mammal kinds (informal but accurate taxonomy with brilliant and unfallen mind). He discovered that no animal was a "helper suitable for him." Eve was created from Adam, thus securing the unity of mankind (Acts 17:26; 1 Cor. 15:22). God performed the first wedding (cf. Matt. 19:6) The universe was now perfect and complete.

From Creation to the Fall

The period from the end of Creation Week to the Fall probably lasted only a few days because (1) the sinless and physically flawless pair were commanded to "be fruitful and multiply" (Gen. 1:28), but no conception took place until after their expulsion from the Garden; and (2) it seems inappropriate for Adam and Eve to have continued long in a state of unconfirmed righteousness when the very purpose of their creation was to glorify God by worshiping Him

Third Day A vast land area (or areas) suddenly rose above the sea level at the word of the Son of God. He placed upon it all the basic kinds (Hebrew *min,* cf. Lev. 11:13–22) of grasses, herbs, and fruit trees, all "full-grown" like Adam (but with no unnecessary marks of age such as growth rings) and with the seed "in itself" for perpetual reproduction, each "after its kind." He had no difficulty sustaining them one day before the sun was created (cf. Rev. 21:23).	**Fourth Day** The sun, moon, and stars (including planets, etc.) were "made" (synonymous with "created" in this chapter as evidenced by comparing 1:21 with 1:25, and 1:26 with 1:27) three days after the earth, thus demolishing the validity of sun worship (cf. Job 31:26–28; Ezek. 8:16). The pseudoscience of astrology is totally rejected by God (Isa. 47:13; Jer. 10:2). They were designed for *illumination* (with light rays created together with light sources in a fully functioning universe), for a *clock/calendar system*, and for *signs* (Ps. 19:1–6; Rom. 1:20).
Seventh Day God was not exhausted from His work of creation (cf. Isa. 40:28–31), but He set apart the seventh day ("sanctified it") and blessed it for man's benefit that he might honor and worship Him in a special way. It was made legally binding only for Israel (Exod. 20:9–10; Col. 2:16). The creation sabbath does not continue to this day, for otherwise God would have cursed the very day He blessed. It lasted twenty-four hours like the other six.	

voluntarily in conscious rejection of sinful alternatives (cf. John 4:23–24; 1 John 2:15).

The Fall (1) brought immediate spiritual death to Adam and Eve, (2) made them physically mortal, (3) changed Eve's body so that she would bring forth children in pain, (4) initiated carnivorous characteristics in many animals, and (5) transformed agriculture from a delight to a drudgery. See *The Genesis Flood,* pp. 454–73.

the sun, moon, and earth at all. When the stupendous fact begins to dawn upon us, however, that these great astronomic bodies were created *instantaneously* and *ex nihilo*, all serious questions and doubts concerning the deity, power, and glory of the Creator evaporate (cf. Rom. 1:20). This is the reason why the Hebrew/Christian approach to creation is not only shocking but also transforming in its impact on the human mind.[43]

Creation Involved a Superficial Appearance of History

The supernaturalism and suddenness of creation provide a necessary background for the concept of creation with a superficial appearance of history or age. Few doctrines of Scripture have met with such misrepresentation and ridicule as this, not only by secular writers but also by some who claim to be evangelical Christians. At the same time, however, few doctrines are quite as far-reaching in their theological significance, and that for at least two reasons.

In the first place, if this doctrine were not true, there could have been no original creation by God at all. Henry M. Morris has made this point quite clear: "If God actually created anything at all, even the simplest atoms, those atoms or other creations would necessarily have an appearance of *some* age. There could be no *genuine* creation of any kind, without an initial appearance of age inherent in it. It would still be possible to interpret the newly-created matter in terms of some kind of previous evolutionary history. And if God could create atomic stuff with an appearance of age—*in other words, if God exists!*—then there is no reason why He could not, in full conformity with His character of truth, create a whole universe full-grown."[44]

In the second place, if the doctrine of creation with appearance of history is erroneous, then most of the recorded miracles of the Lord

43. Even with regard to the creation of living things, it is more appropriate to speak of "instantaneous creation" than "creative process," for our Lord's sign-miracle healings can hardly be described as "processes." Meredith G. Kline attempted to find providential process in Genesis 2:5 ("Because It Had Not Rained," *Westminster Theological Journal* 20:2 [May, 1958], pp. 146–57), but Edward J. Young demonstrated the invalidity of this appeal and insisted that a sharp a line of demarcation exists between "special, divine, creative fiats" and God's normal work of providence (*Studies in Genesis One*, pp. 58–65. Carl F. H. Henry has also rejected Kline's view (*God, Revelation and Authority* [Waco, TX: Word Books, 1983], 6:134).

44. John C. Whitcomb, and Henry M. Morris, *The Genesis Flood* (Phillipsburg, N.J.: Presbyterian and Reformed Pub. Co., 1961), p. 238.

Jesus Christ could not have occurred. One evening on a mountainside near the Sea of Galilee, five thousand men and their families ate loaves and fishes that were created with an appearance of age. Here were tens of thousands of barley loaves composed of grains that had neither been harvested from fields nor baked in ovens! And here were at least ten thousand fishes that had never hatched from eggs or been caught in nets or been dried in the sun!

> The miracles of the feeding of the 4,000 and of the 5,000 involving the instantaneous creation of both animal and plant material certainly casts light on the creation of the animals and plants in Days Three, Five, and Six of Genesis 1.[45]

An equally clear example of this is recorded in the second chapter of John. When Christ began His public ministry on earth, the very first miracle He performed "manifested His glory" (John 2:11) as the Creator of the world (1:3, 14). How did He accomplish this? By instantly transforming about 150 gallons of water into delicious wine. Now wine is the end product of a long series of complex natural processes involving the drawing of water from the soil into the fruit of the grapevine, and the gradual transforming of this water into the juice of grapes. Even then, the ripened grapes must be picked, the juice squeezed out and the sediments allowed to settle down. But Jesus, the Lord of Creation, bypassed all these natural and human processes and created the end product with an appearance of history.

> The fact that our Lord started with water should not detract from the fact that it was a genuine miracle of creation. Our Lord took H_2O and turned it into $C_6H_{12}O_6$ (fructose, the sugar found in wine), as well as the many other products found in wine. There was not only the direct creation of billions of carbon atoms, but also the arranging of all of these atoms into the highly complex molecules in wine. None can deny that it was sudden.[46]

Now it is instructive to note that the headwaiter who "did not know where it came from," naturally assumed that this "good wine" had been somewhere "kept . . . until now" (2:10). This was a natural conclusion, of

45. Marvin L. Lubenow, "*From Fish to Gish*": *Morris and Gish Confront the Evolutionary Establishment* (San Diego: CLP Publishers, 1983), p. 196.

46. Ibid.

course, for neither he nor anyone else in the world had ever considered the possibility of wine coming directly from water. It *must*, therefore, have had a history of natural development. But he was mistaken. He did not know about the supernatural powers of Christ, the Creator God. If I understand the Scriptures correctly at this point, this is the underlying reason for *all* denials of supernatural creation. When contemplating the created works of the Lord Jesus Christ, whether it be sun, moon, earth, oceans, plants, animals, or human beings, the natural man, like the ruler of the feast, simply assumes that they have all been "kept" somewhere "until now," having passed through complex natural processes, from simple primitive forms, through vast periods of time.

It is not difficult to see how this principle applies also to every great miracle of healing performed by our Lord. The ninth chapter of John tells of a man born blind to whom Jesus gave perfect vision. The rulers of Israel refused to believe that the man brought before them could have had a *past history of congenital blindness*—until they consulted his parents. Their perplexity is understandable. As the healed man himself expressed it, "Since the beginning of time it has never been heard that anyone opened the eyes of a person born blind" (9:32). In a moment, Jesus created the appearance of a man born with normal eyesight.

In like manner, Jesus created in Lazarus of Bethany the appearance of a man who had not yet died. Who, in his "right mind," would have imagined that the recent history of this man sitting at a table in Bethany (John 12:2) included four days of decomposition in a tomb? "Because the decay process involves the breakdown of complex biological compounds into simple ones, every cell in the body of Lazarus had to be recreated and restored to its original complexity. That it also was sudden none can deny."[47] Every instance, therefore, of supernatural, sudden, and perfect healings of sick, crippled, or dead people involved the creation of the appearance of an immediately previous condition of health and strength that had not existed. Every priest who was called upon to examine the lepers whom Jesus cleansed must have pondered this question (cf. Matt. 8:4).

Modern critics of this doctrine frequently identify biblical creationism with the extreme views of Philip Henry Gosse (1810–88), who wrote a

47. Ibid.

book entitled *Omphalos: An Attempt to Untie the Geological Knot*.[48] Gosse not only believed that Adam was created with a navel (hence the name of the book, from the Greek word for navel), but also that all conceivable geologic formations, including fossiliferous strata, were created *in situ*.[49]

G. J. Renier comments, "If Philip Gosse is right, a fundamentalist Christian can be a scientist, but it is impossible for him to be an historian."[50] With this judgment we are in general agreement, for Gosse's concept of the creation of fossils actually involves a *denial of biblical history*, especially the history of the Edenic Curse with the introduction of physical death as an effect of man's sin, and the great Flood with its unique capacity for the rapid burial of plants and animals in stratified formations.

Furthermore, the Bible does *not* imply that Adam had a navel, for the lack of this mark of fetal connection with a mother would hardly have constituted Adam an imperfect being. By the same token, the first trees did *not* necessarily have growth rings within them, unless it can be shown that these would be essential to the life of a tree. We may be assured that God did *not* create a world filled with unmistakable and essentially unnecessary testimonies to a previous history simply for the purpose of deceiving men! This is why I prefer to use the expression "superficial appearance of age" to describe the original creation. In the last analysis, however, Scripture alone must serve as our guide for determining what God actually did create, in the biblical sense of that term.

In Genesis 1:11 God commanded that the earth to "'sprout vegetation, plants yielding seed, and fruit trees bearing fruit after their kind, with seed in them, on the earth.'" How are we to understand this? For a number of years I would have agreed with those who insist that here, at least, we have biblical evidence for process in creation.[51] However, further study of the biblical text has led me to abandon this position. The proper

48. *Omphalos: An Attempt to Untie the Geological Knot* (London: John van Voorst, 1857). Cf. Loretta Rouster, "Father and Son: The Tragedy of Edmund Gosse" (*Creation Social Science and Humanities Quarterly* 2:3 [Spring, 1980] 10–12; and David Pott, "Philip Henry Gosse (1810–1888)," *Biblical Creation* 4:13 (January, 1983) 75–83.

49. Ibid., p. 347.

50. *History: Its Purpose and Method* (New York: Harper Torchbooks, 1965), p. 126.

51. Cf. Russell Mixter, *Evolution and Christian Thought Today* (Grand Rapids: Eerdmans, 1959), pp. 69, 151.

context for understanding the events of creation week is not our present world of noncreative process (the first and second laws of thermodynamics), but rather the person and work of the Lord Jesus Christ as unveiled in the New Testament. If nearly every visible miracle performed by our Lord on earth involved the creation of built-in history, should we expect anything less during that unique period when He brought the world into existence? When He commanded the earth to bring forth fruit trees, did He have to create seeds first and then wait a number of years for them to grow to maturity? It would be much more in harmony with His later works in the Holy Land to understand this command as being fulfilled by a sudden appearance of full-grown fruit trees bearing fruit. Doubtless it will be objected that this is contrary to God's normal and observable way of bringing fruit trees into existence today. Quite true! But if this line of argument were pursued consistently, then God could not have created the first seeds either, for if natural observation be our guide, the highly complex seeds of fruit trees can only come from fruit trees.

The Old Testament itself provides important analogies for the supernatural creation of "full-grown" vegetation. Note, for example, God's description of Aaron's rod only a few hours after it had been placed as a dead stick in the Tabernacle: ". . . and behold, the rod of Aaron for the house of Levi had sprouted and put forth buds and produced blossoms, and it bore ripe almonds" (Num. 17:8). For centuries this miracle object remained in the ark "as a sign" (17:10). Compare the verbs used here ("sprouted," "put forth," "produced," and "bore") with the verbs in Genesis 1:11 ("sprout," "yielding," "bearing"). One must also ponder the significance of the large shade-plant which "came up overnight" east of Nineveh by God's miraculous working on behalf of His distraught prophet (Jonah 4:6–10). Surely God's original creation of the plant kingdom deserves to be remembered, along with Aaron's rod and Jonah's shade-plant, as a sign of his *supernatural* power and wisdom.

It is really quite impossible to escape the conclusion that if God created living things *after their kind*, as the first chapter of Genesis states ten different times, He must have created them with a superficial appearance of age. And the Scriptures inform us that God began the cycle of life with adult organisms rather than with embryonic forms. Both Old and New Testaments concur in the supernatural creation of Adam and Eve, as adults. And must not this have been true also of all the kinds of animals? How could such creatures have existed as mere fertilized eggs outside of the mother's womb? And how could infant mammals have survived without a mother's care? God would have had to intervene directly and continually to care for them. Therefore, unless we appeal to an endless supplying of miracles, the direct creation of *adult* organisms remains as the only logical interpretation of the Genesis account of the creation of living things *after their kind*.

Several years ago Thomas H. Leith opposed this concept in a paper entitled, "Some Logical Problems With the Thesis of Apparent Age," which he presented at the Nineteenth Annual Convention of the American Scientific Affiliation.[52] In the first place, Dr. Leith claimed that such a doctrine lacks empirical evidence and undermines all true science. But if this is true, then all miracles in the Bible can be denied, for on the same basis it could be claimed that the virgin birth of Christ lacks empirical evidence and undermines the sciences of genetics and biology. He dismissed the raising of Lazarus as an analogy for creation with apparent history, for in this case, he claims, human observers were present to see the miracle, whereas the supposed discontinuities of Genesis (such as Creation and the Flood) were not observed! Presumably for Dr. Leith the Book of Genesis is not as historically dependable as the Gospel of John, or at least the creative acts recorded in the first two chapters of Genesis do not fit proper standards for empirical verification because human observers were not present to study them. In other words, he seemed to be implying that God is not a trustworthy witness of what happened at the time of creation.

Leith's second main objection to the doctrine of apparent age was that it makes God a deceiver of men. "One wonders," he asked, "why deity should be so malevolent (like a Cartesian demon) as to fool us on such interesting matters as much of the history of past events and the possible ages of many things, especially when it is the sort of delusion from which we poor mortals cannot escape!"[53] To this rather common objection, we need only reply that God has not deceived us in such matters if He has given us an infallible Book to tell us what He has done. We have no one to blame but ourselves if we reject the written record of His creative and miraculous works in history.

Edward J. Carnell suggested two principles to guide us in this area: (1) "since we have God's promise to sustain a regular universe, a Christian

52. *Journal of the American Scientific Affiliation* 17:4 (December 1965), pp. 118 ff.
53. Ibid., p. 122.

Fruit Trees

Fruit trees are not the end products of a billion years of evolutionary development from marine protozoans. Instead, they were created by God two days before any marine life appeared, along with all other kinds of plants. Neither did they grow from seeds, but were created full-grown (without growth rings). God did create seeds, but they were inside of fruits hanging from full-grown trees ("and fruit trees bearing fruit after their kind, with seed in them" Gen. 1:11). Similarly, the first human beings neither evolved from ape-like ancestors nor did they grow up from fertilized eggs or babies. They were created full-grown (without navels), fully capable of obeying God's command to "be fruitful and multiply, and fill the earth" (Gen. 1:28). The only concept of origins that truly fits the observable facts of genetics and paleontology is supernatural creationism, whereby vast numbers of distinct and unique kinds of living things suddenly appeared with the capacity of reproducing after their kind. Compromise positions, such as theistic evolution, neither fit the clear statements of Scripture nor the discoveries of empirical science.

could defend the principle of uniformity until he either falls into absurdity or departs from Scripture; (2) we must cheerfully admit God's moral right to create things which only appear, but are not actually, old. The limits of how God has employed this privilege must be measured—in the last analysis—not from science, but from Scripture." I would concur with his final conclusion: "These two principles may be hard to apply. True. But there is one thing which is much harder, and that is to rescue Christianity from the jaws of science once the principle of uniformity destroys God's right to perform miracles."[54]

If Scripture be our standard in all truth, then creation with appearance of age is not deceptive, but glorious. Did Jesus deceive the ruler of the feast when He changed water to wine? The Word of God provides the answer for us: "This beginning of His signs Jesus did in Cana of Galilee, and *manifested* His *glory*" (John 2:11). The glory of Christ was revealed in this miracle because it involved the supernatural and sudden creation of a complex entity apart from natural processes. And this, we believe, is how the Lord Jesus Christ revealed His glory in the creation of the world.

Conclusion

The absolute sovereignty and wisdom and power of God, and thus His glory, are overwhelmingly revealed in the method and timing of His creative work. General revelation, though full of testimony to His greatness (Ps. 19:1–4), cannot speak to the heart of man with the precision and urgency that characterizes His written revelation. This is especially true of the opening words of Genesis; for here we find, in the creation time-block of six literal days of creation, a grid through which the face of God, in the Person of His Son Jesus Christ, can be seen especially in New

54. "Beware of the New Deism," *His Magazine*, 12:3 (December 1951), pp. 14ff. See the answer to Thomas Leith by Lloyd G. Multhauf, Department of Physics, Pennsylvania State University, *Journal of the American Scientific Affiliation*, 18:2 (June 1966), p. 63. See also the helpful discussion of mature creation in Donald E. Chittick, *The Controversy: Roots of the Creation-Evolution Conflict* (Portland, OR: Multnomah Press, 1984), pp. 196ff. Dr. Chittick is an experienced scientist and teacher, and is convinced that evolutionism is actually "anti-science" (p. 116). Henri Blocher, on the other hand, is a theologian who has been so deeply influenced by "majority opinions" that are "currently in favor amongst scientists," that he can boldly label the many dedicated creationists in the scientific community as "anti-scientists" (*In the Beginning* [Downers Grove, IL: InterVarsity Press, 1984], p. 241; cf pp. 213–31). Clearly, one's presuppositions concerning the nature of biblical truth are of vital importance in viewing both ultimate origins and empirical science.

Testament retrospect (John 1:1–3). Just as He performed his creative miracles during the days of His non-glorified incarnate ministry in Palestine, so also during creation week, He did His divine work suddenly, supernaturally, and with a superficial appearance of 'history.' The theory of evolution has effectively obscured this marvelous truth, even among many of God's people, and therefore must be removed from our hearts through a renewed submission to the Holy Spirit who gave us the Bible through chosen men (2 Pet. 1:19–21). Only in this way can we avoid being "conformed to this world" in our understanding of ultimate origins, and finally "prove what the will of God is" in our handling of the intricate and precious details of biblical creationism.

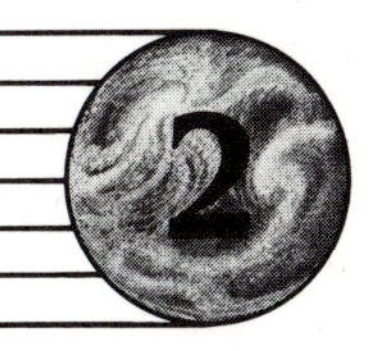

The Creation of the Universe

The Basic Approach to Origins

We have seen that the Word of God teaches the supernatural and instantaneous creation of all things. With regard to physical entities in particular, we may add the concept that no preexistent materials were used. In the strictest sense, this is the meaning of Hebrews 11:3—"By faith we understand that the worlds [*aiōnas*—literally, "ages," the universe of mass/energy in its time extension] were prepared by the word of God, so that what is seen was not made out of things which are visible" (cf. Rom. 4:17). This certainly cannot mean that the physical substances that compose our visible universe consist of "invisible" atomic particles! Spiritual faith is certainly not required to accept the atomic theory of matter in its current form! The point of this key verse on creationism is that visible material substances did not exist in any form whatsoever, other than in the mind of an omniscient God, until He spoke the creative word.

The informed and spiritually-minded Christian frankly acknowledges, in conformity with the clear statement of Hebrews 11:3, that his understanding of the order of events and the methods employed by the

Creator in bringing the world into existence is basically a *faith-commitment* to God's special revelation. It is "by faith," not by empirical observation, that he "understands" the *ex nihilo* approach to ultimate origins. And his confidence in the absolute authority and dependability of God's written revelation in the Bible is based, in turn, upon a profound assurance that his Lord, Jesus Christ, who put His divine imprimatur upon the Scriptures, was neither deceived nor a deceiver, but spoke final truth (cf. John 14:6; Matt. 5:18; John 5:46).

At the same time, and in complete honesty, the non-Christian scientist must acknowledge that *he also* comes to the factual, observable phenomena with a set of basic assumptions and presuppositions that reflect a profound "faith-commitment." No scientist in the world today was present when the earth came into existence, nor do any of us have the privilege of watching worlds being created today! Therefore, the testimony of an honest evolutionist could be expressed in terms of the same Hebrews 11:3 outline, as follows: "By faith, I, an evolutionist, understand that the worlds were *not* framed by the word of any god, so that what is seen has indeed been made out of previously existing and less complex visible things, by purely natural processes, through billions of years."

Thus, it is not a matter of the *facts* of science versus the *faith* of Christians! The fundamental issue, in the matter of ultimate origins, is whether one puts his trust in the written Word of the personal and living God who *was* there when it all happened, or else puts his trust in the ability of the human intellect, unaided by divine revelation, to extrapolate presently observed processes of nature into the eternal past (and future). *Which faith* is the most reasonable, fruitful, and satisfying? In my own case, while studying historical geology and paleontology at Princeton University I was totally committed to evolutionary perspectives. Since then, however, I have discovered the biblical concept of ultimate origins to be far more satisfying in every respect.

Christians who truly desire to honor God in their thinking must not come to the first chapter of Genesis with preconceived ideas of what could or could not have happened (in terms of current and changing concepts of uniformitarian scientism). We are not God's counselors; He is ours! "FOR WHO HAS KNOWN THE MIND OF THE LORD, OR WHO BECAME HIS COUNSELOR?" (Rom. 11:34) ". . . 'For My thoughts are not your thoughts, neither are your ways My ways,' declares the LORD. 'For as the heavens are higher than the earth, so are My ways higher than your ways, and My thoughts than your thoughts'" (Isa. 55:8–9).

The Creation of the Heavens

For convenience of human thought and expression, the Bible refers to three different heavens. The *third* heaven is that glorious place surrounding the immediate presence of God, to which Paul was carried in a transcendent vision early in his Christian experience (2 Cor. 12:1–4). The *second* heaven seems to be equivalent to what we call "outer space"; while the *first* heaven consists of the atmospheric blanket surrounding the earth, in which clouds move and birds fly.

In the first chapter of Genesis, a distinction may be seen between the first heaven, above which the waters were lifted (vss. 7–8, 20) and the second heaven in which the luminaries were placed (vss. 14–17). There is certainly nothing crude or "prescientific," in the bad sense of that expression, about the cosmology of Genesis, as many able expositors have successfully and repeatedly demonstrated.[1]

What were the "heavens" like at the moment they came from the Creator's hand "in the beginning"? The *third* heaven was populated with hundreds of millions of angelic beings (Dan. 7:10; Rev. 5:11; 9:16), each one a "son of God" in the sense of a direct creation by God (cf. Job 1:6) and therefore perfect in all their ways (cf. Ezek. 28:15). They must have been created at the very beginning of the first day of creation, for Job 38:6, 7 tells of their singing and of their shout of joy at the creation of the earth.

That they did not exist *before* the first day is indicated by Colossians 1:16 (which tells us that Christ created all *invisible* as well as visible thrones, dominions, principalities and powers *in the heavens* as well as upon the earth) in the light of Exodus 20:11 ("For *in six days* the LORD made the heavens and the earth, the sea, and *all that is in them*").

The *second* heaven, the realm of "outer space," was presumably empty and dark, for the sun, moon, and stars were not created until the fourth day, and the special light source which divided the light from the darkness had not yet been spoken into existence.

The *first* heaven, or atmospheric blanket, had neither vapor canopy nor clouds, for the waters were not yet lifted above the expanse ("firmament") in the form of a vast, invisible thermal vapor blanket, as must have existed until the Flood, and there were no clouds or rain as in our present post-

---

1. Cf. R. Laird Harris, "The Bible and Cosmology," *Bulletin of the Evangelical Theological Society*, 5:1 (March, 1962), pp. 11–17.

The Oceans

In all the universe, Planet Earth is the only known place where liquid water exists; and here there are 330,000,000 cubic miles of it!

> It pours down on the earth at the rate of 1.5 trillion tons a day. It covers 72 percent of our planet's surface, 70 billion gallons for every person alive. . . . Water is important to our existence and is found to be delicately balanced in all of its physical properties for our benefit. The phrase "water of life" is also found in Revelation 22:17 with reference to salvation. How appropriate that water, the most mentioned natural resource in the Bible, should be used to symbolize the Creator's greatest gift to His creatures. Both are free; both are priceless (Donald B. DeYoung, "The Water of Life," *Creation Research Society Quarterly* 22:3 [December, 1985], pp. 107–14).

The ocean basins of our present world, since the Flood, are much deeper than those before the Flood, because they now serve as reservoirs for "the waters which were above the expanse" as well as "the waters which were below the expanse" (Gen. 1:7). In fact, whereas Mt. Everest is 29,028 feet (8,848 meters) above sea level, the deepest ocean (Mariana Trench near Guam in the Pacific) is 35,810 feet (10,915 meters) deep! When "the floodgates of the sky were opened" by God at the beginning of the Flood year, the vast vapor canopy condensed and collapsed in the form of torrential rains within six weeks (Gen. 7:11–12). At the end of the Flood year, "the valleys [basins] sank down," and these great masses of water which "were standing above the mountains" now "fled" and "hurried away . . . . To the place which Thou didst establish for them. Thou didst set a boundary [the shorelines of the world] that they may not pass over; that they may not return to cover the earth" (Ps. 104:6–9). This great rainbow covenant (cf. Gen. 9:8–17; Isa. 54:9) is our guarantee that the oceans have reached their final resting place. When the present earth is replaced by a new earth, there will be "no longer any sea" (Rev. 21:1).

Flood world.[2] Neither Genesis nor geology gives any support to the idea that earth's primitive atmosphere consisted of ammonia, methane, hydrogen, and water, as the evolutionary theory of spontaneous generation of life requires (see chapter 3).

Some Bible students believe that the heavenly bodies were created in the beginning, but could not be seen from the earth because of a cloud blanket so dense that darkness covered the face of the deep. However, the waters were not lifted up until the second day, and the light that was created on the first day was clearly visible from the earth. Also, if God's work on the fourth day involved merely the unveiling of previously created heavenly bodies, this idea could have been more clearly expressed by the use of the verb "appear" as in verse 9—"and let the dry land *appear*." Instead of this, we are told that God "made" two great lights on the fourth day, and that He "made" the stars also.[3]

The Creation of the Earth

The earth, like the heavens, was created without the use of preexistent materials (Heb. 11:3), which clearly implies that it was created instantaneously as a dynamic, highly complex entity. It was spinning on its axis, for in reference to the localized light source created on the first day (Gen. 1:3), it passed through three night/day cycles. It had a cool crust, for it was covered with liquid water.

The crust, however, had no significant features, such as continents, mountains, and ocean basins, for these were formed on the third day. Nor did it have sedimentary strata or fossils, for these were basically the effects of the Great Deluge. But it did contain all of the basic elements and the foundational rocks of our present earth. As a planet, it was perfect in every way, but at this stage of creation week it was not yet an appropriate home for man. It was "formless and void" or "unformed and unfilled" *tōhû wāḇōhû*. (See chapter 5 for a discussion of the Gap Theory of Genesis 1:2.)

God, of course, could have filled the earth with living creatures on the first day; but Exodus 20:11 suggests that He did it in six days in order to provide a glorious pattern for Israel's work week. Therefore, we must not

2. See Joseph C. Dillow, *The Waters Above: Earth's Pre-Flood Vapor Canopy*, rev. ed. (Chicago: Moody Press, 1982), and the review by J. C. Whitcomb and D. B. DeYoung in *Grace Theological Journal* 3:1 (Spring, 1982). pp. 123–32.

3. See Chapter 3 for a discussion of the words "made" and "created."

judge the quality of God's creative work by the appearance of the earth at the end of the first day. It was merely the first of six twenty-four-hour stages of creation.

Did the Earth Come from a Proto-sun?

If Genesis teaches that the earth was created *before* the sun, moon, and stars, then Christians who believe the Book of Genesis are obviously in serious conflict with evolutionary theory at this point. For this reason, many Christians feel that Genesis must be interpreted in such a way as to avoid this conflict. After all, they reason, is it not perfectly clear from astronomical studies that the earth and the other planets came from the sun or from a proto-sun, and that, in turn, from a 'big bang'? It shall be our purpose in the following paragraphs to show that this is not true.

Early in this century, catastrophic models for the solar system's origin were popular. T. C. Chamberlain and F. R. Moulton envisioned the close encounter of another star with the sun. The resulting tidal effects supposedly drew off embryonic planets. However, these men couldn't explain how hot stellar material could condense into planets instead of dissipating. Also, their Collision Theory became unpopular because the trillions of miles that separate the stars would make planet formation by this method exceedingly rare if not impossible.

In recent years astronomers have returned to a modified form of the old "nebular hypothesis." They propose the collapse of an interstellar gas cloud, in stages. First, dust grains supposedly collided and stuck together, building up fist-sized balls. These clumps then bunched up or snow-balled in an "accretion" process, growing over millions of years into separate planets and moons. Recent proponents of this theory include A. G. W. Cameron, T. Gold, W. Hartmann, and P. Goldreich.

How well this currently popular theory succeeds in explaining the solar system in terms of physical, chemical, and mathematical principles alone may be decided only after considering nine of the basic problems which remain to be solved by evolutionary cosmogonists.

1. Before any condensation of gas and dust could occur, the nebula would have diffused into space. J. A. Wood states,

> Planetary accretion, like most other aspects of solar system origin, is imperfectly understood. Once planetary nuclei (objects some tens of kilometers in dimension, say) had gotten started, it is easy enough to see how they would grow by sweeping up smaller particles. But it has always

> been difficult to see how the start was made; why dust particles, chondrules, and Ca, Al-rich inclusions chose to clump together.[4]

Furthermore, why don't we see this happening within the ring systems of Saturn, Jupiter, Uranus, or the asteroid belt? A popular idea today is that the original gas cloud was compressed by the pressure wave from a nearby exploding star, a supernova. The problem is that no nearby supernova remnant has been located.

2. The theory demands a complex system of "roller-bearing eddies" of gas and dust for the gas giants (i.e., Jupiter, Saturn, Uranus, Neptune). However, this is impossible because such vortices must remain perfectly intact during essentially the entire period of planetary accretion. Gerard P. Kuiper conceded, "It is difficult to conceive that the beautiful system of vortices would actually have been in existence long enough—even for 10 or 100 years—to get the condensation of the building material for the planets under way."[5] Yet the theory demands many *millions* of years.

3. What stopped the process from continuing so that the entire mass of material did not form one large body? The sun makes up 99 and 6/7 percent of the mass of the solar system, so what would have kept the remaining 1/7 of one percent from falling into the sun?

4. There is much interstellar material in the vicinity of the sun, but it is not condensing. Greenstein of the Mount Wilson Observatory believed that the known stars rotate so fast they they could never have been formed by a condensation process. In fact, many stars have a rotation speed one hundred times that of the sun! With this speed, such stars should not be able to hold on to their surface layers. But if this is happening, how did such stars collapse in the first place? The initial gas clouds should have developed a stable circulating motion without collapsing into stars. Indeed, flat disks of gas material have been observed around nearby stars such as Vega and Beta Pictoris, but whether this gas is contracting or dissipating is entirely unknown.

5. The planets contain less than 1 percent of the mass of the solar system but a staggering 98 percent of its angular momentum. Jupiter itself possesses 60 percent of the solar system's entire angular motion. This distribution was the main failure of the old nebular hypothesis. The solar material, as it collapsed inward, should have spun the sun up to a

4. J. A. Wood, *The Solar System* (Prentice-Hall, Inc. 1979), p. 167.

5. Gerard P. Kuiper in *Astrophysics: A Topical Symposium*, ed. J. A. Hynek (New York: McGraw-Hill, 1951).

high rotational speed. Recent explanations of the sun's slow motion appeal to "magnetic braking." This proposed interaction between the solar magnetic field and charged nebular particles remains speculative.

David Layzer, professor of astronomy at Harvard University, found no solution to the problem of the small angular momentum of the sun. If it had been part of a gaseous protogalaxy, its angular momentum would have to be a *billion* times as much as it now possesses. How it could have lost all but *one ten millionth of one percent* of its original angular momentum has never been explained.[6]

6. The evolutionary accretion theory cannot easily explain why seven of the nine planets have direct rotation in reference to their revolution around the sun, but Venus rotates slowly backwards, and Uranus rotates at a 98-degree angle from its orbital plane, even though its orbit inclines less than that of any other planet. In general, catastrophic collisions between solar system objects are thought to account for such anomalies. But detailed explanations are lacking. And if major collisions were frequent, why is there such a high degree of order in the solar system? From the perspective of one evolutionist, even though Voyager 2 passed close to Uranus in January, 1986, "The spacecraft's fabulous set of data [did not] shed any clear light on why a planet should evolve as Uranus did, spinning so oddly. Perhaps . . ., despite everything found in January, we'll never know the answer."[7]

7. Cosmic evolutionism cannot explain retrograde satellites (backward-orbiting moons). Of the fifty-two moons in our solar system, *twenty-one* are orbiting backwards relative to the direction of orbit of the nine planets around the sun.[8] Of special interest is Triton, the inner of Neptune's two satellites, which has nearly twice the mass of our moon (its diameter being 3,000 miles) and which revolves *backwards* every six days in a nearly circular orbit only 220,000 miles from Neptune (closer than our moon to the earth)!

Isaac Asimov, as well as most evolutionary cosmogonists, believes that Triton "was thrown away from that planet by some cosmic collision or other accident," and that later on Neptune recaptured its lost moon into a retrograde orbit by "a similar accident."[9] But how many such

6. David Layzer, "Cosmogony," in *McGraw-Hill Encyclopedia of Science and Technology*, 15 vols. (New York: McGraw-Hill, 4th ed., 1977), III, 561.

7. J. Kelly Beatty, "A Place Called Uranus," *Sky and Telescope* (April, 1986), p. 337.

8. For a listing of the ten newly discovered (retrograde) moons orbiting close to Uranus, see *Sky and Telescope* (April, 1986), p. 341.

9. *The Intelligent Man's Guide to Science*, 2 vol. (New York: Basic Books, Inc., 1960) I:78.

The Great Galaxy in Andromeda

"He made the stars also" (Gen. 1:16). This gigantic spiral galaxy composed of scores of billions of individual stars is the only one outside of our own Milky Way that is visible to the unaided eye in the Northern Hemisphere. It is two million light years (ten quintillion miles!) away from the earth. Nevertheless, its light rays were created by God already reaching the earth (Gen. 1:15) so that its God-intended function of serving as one of the heavenly "signs" of God's glory and handiwork (Gen. 1:14, Ps. 19:1) and as one of the "clearly seen" testimonies to His eternal power and divinity (Rom. 1:20) could be effectively accomplished. God did not have to create this galaxy two million years beforehand in order for its light to have sufficient time to reach the uplifted eyes of our first human parents. See Paul M. Steidl, *The Earth, The Stars, and the Bible* (Phillipsburg, NJ: Presbyterian and Reformed Publishing Company, 1979), pp. 219–24.

"accidents" may one be permitted to invoke to prop up a theory already tottering under the weight of its own unproved assumptions? Asimov further states that retrograde satellites are "minor exceptions" to the general rule of satellite orbits.[10] However, twenty-one out of fifty-two moons having retrograde orbits can hardly be brushed aside as "minor exceptions."

8. What can evolution really offer as an explanation of the angular momentum in these satellite systems? We will permit Professor Layzer of Harvard to state the problem:

> Except in the Earth-Moon system (which is exceptional in other respects as well), the primary carries the bulk of the angular momentum, instead of the satellites. . . . This circumstance aggravates the theoretical difficulty presented by the slow rotation of the Sun, for if the Sun has somehow managed to get rid of the angular momentum it would be expected to have, according to the nebular hypotheses, why have the planets not done likewise?[11]

9. In spite of some ingenious and very complicated theories, it has never satisfactorily been shown why the earth is composed of such heavy elements. In the words of Professor Fred Hoyle of Cambridge University:

> Apart from hydrogen and helium, all other elements are extremely rare, all over the universe. In the sun they amount to only about 1% of the total mass. . . . The contrast [with the heavy elements which predominate in the earth] brings out two important points. First, we see that material torn from the sun would not be at all suitable for the formation of the planets as we know them. Its composition would be hopelessly wrong. And our second point in this contrast is that it is the sun that is normal and the earth that is the freak. The interstellar gas and most of the stars are composed of material like the sun, not like the earth. You must understand that, cosmically speaking, the room you are now sitting in is made of the wrong stuff. You yourself are a rarity. You are a cosmic collector's piece.[12]

In the light of such considerations, one prominent astronomer concluded: "I think that all suggested accounts of the origin of the Solar

10. *Asimov's New Guide to Science* (New York: Basic Books, Inc., 1984), p. 97.

11. "Cosmogony," *McGraw-Hill Encyclopedia of Science and Technology*, III, 564.

12. *Harper's Magazine* (April, 1951), p. 64. Quoted by Paul Zimmerman in *Concordia Theological Monthly* 24:7 (July, 1953), p. 506.

System are subject to serious objections. The conclusion in the present state of the subject would be that the system cannot exist."[13] Whipple concurs: "All of the hypotheses [regarding solar system formation] so far presented have failed, or remain unproved, when physical theory is properly applied."[14]

The overwhelming evidence of design throughout the entire universe as well as in the solar system and our own planet has never been more obvious than now. The perfect mass of the proton, and the exact factor of 2 in gravitational and electrical force equations demand a supreme Designer. So remarkable are these universal mathematical proportions that the term Anthropic [Human] Principle is being widely used among astronomers to describe the "neat and tidy" cosmic mathematical formulas which are independent of the human mind and yet seem to be in beautiful harmony with the way we think.[15] The more we learn of the astronomic universe, the more we realize that evolutionism, even "theistic evolutionism," offers no rational answers.[16]

Thus, in our generation, even more than in ancient times, God's "eternal power and divine nature have been clearly seen." These attributes should be obvious to mankind because they are "being understood through what has been made, so that they are without excuse" (Rom. 1:20). But men, in their state of rebellion against the God of

13. Sir Harold Jeffries, *The Earth: Its Origin, History and Physical Constitution* (Cambridge, England: University Press, 1970), p. 359.

14. Fred L. Whipple, *Orbiting the Sun* (Cambridge: Harvard University Press, 1981), p. 284. See also the evidences for the supernatural design and recent origin of the solar system presented by Paul M. Steidl, "Planets, Comets, and Asteroids," in George Mulfinger, Jr., ed., *Design and Origins in Astronomy* (Norcross, GA: Creation Research Society Books, 1983), pp. 73–106; and his more comprehensive study, *The Earth, The Stars, and The Bible* (Phillipsburg, NJ: Presbyterian and Reformed Pub. Co., 1979).

15. Cf. Donald B. DeYoung, "Impact No. 149. Design in Nature: The Anthropic Principle," *Acts and Facts* 14:11 (November, 1985). DeYoung calls attention to two articles: B. J. Carr and M. J. Rees, "The Anthropic Principle and the Structure of the Physical World," *Nature* 278 (April 12, 1979), pp. 605–12; and G. Gale, "The Anthropic Principle," *Scientific American* 245 (December, 1981), pp. 154–71.

16. For a theological perspective, see John C. Whitcomb, *The Bible and Astronomy* (Winona Lake, IN: BMH Books, 1984). For a critique of the Big Bang Theory, see Donald B. DeYoung and John C. Whitcomb, "The Origin of the Universe," *Grace Theological Journal* 1:2 (Fall, 1980), 149–61. Sir Fred Hoyle concurs: "I have little hesitation in saying that a sickly pall now hangs over the big bang theory. When a pattern of facts becomes set against a theory, experience shows that the theory rarely recovers" ("The Big Bang Under Attack," *Science Digest* 92 [May, 1984], p. 84).

The Moon

"The lesser light to govern the night" (Gen. 1:16) remains as a "faithful witness in the sky" (Ps. 89:37, NIV) to the creative power of God. This fact was overwhelmingly obvious to David: "When I consider . . . the moon and the stars, which Thou hast ordained; what is man . . . (Ps. 8:3).

At the same time, the moon remains a hopeless enigma to evolutionists, even after several fantastically expensive lunar probes (the Apollo Mission), and prolonged, intensive scrutiny of 843 pounds of moon rocks. *Where did the moon come from?* Celestial mechanics and lunar chemistry show us that it could not have condensed from interstellar dust, or come out of the earth, or have been captured by the earth. Furthermore, the lack of dust on its surface remains unexplained in view of the four-billion-year time scale required by uniformitarian scientism (cf. John C. Whitcomb and Donald B. DeYoung, *The Moon,* pp. 35–51; 94–95).

Michael J. Drake observed: "Although it has been fourteen years since the first lunar samples were returned to Earth by the Apollo 11 Mission, the origin of the moon remains unresolved" ("Geochemical Constraints on the Origin of the Moon," in *Geochimica et Cosmochimica Acta,* Vol. 47 [1983], p. 1759). Nafi Toksoz, a geophysicist at the Massachussets Institute of Technology, concedes, "It's far easier to explain why the moon shouldn't be there than to explain its existence" (quoted by Ben Patrusky, "Where Did the Moon Come From?" in *Science 81* (March, 1981), p. 120).

It is not because of such evolutionary failures that Christians affirm the supernatural creation of the moon. Such a position would be a weak "God of the gaps" approach to theology. The ultimate basis for believing that God directly created the moon out of nothing by His omnipotent Word is divine propositional revelation: "And God made . . . the lesser light to govern the night . . . and God saw that it was good" (Gen. 1:16, 18).

creation, are always everywhere suppressing this truth in spite of the fact that what "is known about God is evident within them; for God made it evident to them" (Rom. 1:18–19). Astronomers, it seems, in their own special ways, find comfort in the fact that many theories are available to them to explain the universe by chance, thinking that the ultimate answer must lie somewhere in these theories. Their deep philosophic/religious faith in cosmic materialism effectively blinds them to the utter inadequacy of each and every theory that has ever been proposed. In this sense, secular scientism is "always learning and never able to come to the knowledge of the truth" (2 Tim. 3:7).

Strangely, theistic evolutionists, who claim to be Christians while at the same time accepting evolution as the expression of God's "strategy" of "creating" the world, are often the slowest to see the contradictions and absurdities of evolutionary theories of the origin of the universe and of the earth/moon system and its surrounding planets and moons. While some non-Christian scientists, such as Sir Fred Hoyle of England, are openly acknowledging the serious contradictions inherent in the Big Bang Theory, various Christian scientists and philosophers (including the present leadership of the 2100-member American Scientific Affiliation) are eloquently promoting cosmic evolutionism.

For example, Charles E. Hummel, the director of faculty ministries for Inter-Varsity Christian Fellowship, sees *no* astronomic/geologic/biologic guidelines in Genesis 1, for "The Genesis 1 account of creation was not intended to teach *how* or *when* God created the universe."[17] A biblical justification for this approach to the opening chapters of Genesis, in Hummel's mind, is the supposed contradiction between the creation accounts in Genesis 1 and 2 (a major presupposition of liberal and neo-orthodox hermeneutics). Thus, "in putting the two accounts side by side the writer must have had a purpose in mind other than describing exactly how God created heaven and earth with its inhabitants, including human beings. Therefore it is both futile and misguided for us to try to determine one precise method of creation from these accounts" (p. 251).

By setting Genesis 2 into conflict with Genesis 1, Charles Hummel not only demonstrates his theological incompetence but also exemplifies the failure of theistic evolutionism in general to come to grips with biblical realities. Genesis 2 is in perfect harmony with the opening

17. Charles E. Hummel, *The Galileo Connection: Resolving Conflicts Between Science and the Bible* (Downers Grove, IL: InterVarsity Press, 1986), p. 246. See below, pp. 116–25, for an additional critique of theistic evolutionism.

chapter, for it focuses the attention of the reader on specific details that would not have been appropriate in the broad spectrum of creation events described in Genesis 1 (just as Genesis 11:1-9 explains in detail how the languages mentioned in Genesis 10:5, 20, and 31 came into existence). Furthermore, the details in Genesis 2 set the stage for the tragedy of the Fall. It is thus an historical restrospect or "flash-back" which was commonly used in ancient Near Eastern documents.

K. A. Kitchen, internationally recognized Old Testament scholar at the University of Liverpool, makes the following comment on Genesis 1 and 2:

> Failure to recognize the complementary nature of the subject distinction between a skeletal outline of *all* creation on the one hand, and the concentration in detail on man and his immediate environment on the other, borders on obscurantism. . . . What is absurd when applied to monumental Near Eastern texts that had *no* prehistory of hands and redactors should not be imposed on Genesis 1 and 2, as it is done by uncritical perpetuation of a nineteenth-century systematization of speculations by eighteenth-century dilettantes lacking, as they did, all knowledge of the forms and usages of Ancient Oriental literature.[18]

Another example of the theological naiveté of Christian evolutionists is Howard J. Van Till, Professor of Physics and Astronomy at Calvin College, Grand Rapids, Michigan. For Van Till, the early chapters of Genesis ". . . were never intended to answer questions about precisely what happened." What, then, does the reader do with the intricate historical and chronological details which saturate the six-days creation account? "In the story, God the Creator is clearly portrayed as performing his creative works within a six-day period and resting on the seventh. What must we make of that chronology? . . . The beginning lies shrouded in mist beyond human memory, and the end will come 'as a thief in the night.' The seven-day chronology that we find in Genesis 1 has no connection with the actual chronology of the Creator's continuous dynamic action in the cosmos. The creation-week motif is a literary

18. K. A. Kitchen, *The Ancient Orient and the Old Testament* (Chicago: InterVarsity Press, 1966), pp. 116–19. On the unity and harmony of Genesis 1 and 2, see also G. Ch. Aalders, *Bible Student's Commentary: Genesis*, translated by William Heynen (Grand Rapids: Zondervan Publishing House, 1981), 1:78–81; and Kenneth L. Barker, "A Response to Historical Grammatical Problems," in Earl D. Radmacher and Robert D. Preus, eds., *Hermeneutics, Inerrancy, and the Bible*, p. 136.

The Sun

"The greater light to govern the day" (Gen. 1:16) began its existence in the middle of creation week. It was not "made" on the fourth day in the sense of being unveiled from a cloud covering, for in the opening chapter of Genesis the verb "made" is used synonymously with "created."

From very ancient times, perhaps from the Fall, men have worshiped this brilliant, inanimate creature (Job 31:26; Deut. 4:19). This was especially true in Egypt, where Israel resided for hundreds of years. When God first presented the Genesis record to His people (through Moses) after their departure from Egypt, they must have been astounded to find that their God did not even name the sun (or the moon) in Genesis 1. Furthermore, they discovered that this supposed deity did not even exist when their great God created plants and trees on the earth's continents! This revelation constituted a devastating blow to the concept that the sun was ultimate, eternal, and indispensable.

A proper recognition of the nature of the sun also serves to destroy cosmic evolutionism. The sun's enormous and continual loss of mass/energy (four million tons a second!) points unmistakably to an original creation when high-level energy and order was built into it by God. Thus, the sun provides a magnificent illustration of the second law of thermodynamics and of the utter bankruptcy of naturalistic theories of its origin (cf. Ps. 102:25–27; Isa. 51:6). In the "new heavens" which God will create some day, mankind will no longer need the light of the sun (Rev. 21:23; 22:5). Thus God, and God alone, is essential for the creation and preservation of the earth and its inhabitants.

device . . . [containing] imaginative illustrations of the way in which God and the Creation are related."[19]

For Howard Van Till, then, as for neo-liberal and neo-orthodox thinkers in general, Genesis does not provide for us *genuine primeval history* at all, but simply "theological" insights (depending, of course, on the perspectives of each individual "theologian"). "Cosmic history is evolutionary in character" (p. 189), not only for the origin of the earth (p. 187), but for "life-forms" as well (p. 188).

Such an approach to Genesis 1-11, consistently applied, destroys the theological and historical credibility of the rest of the Old Testament, which stands upon it as a building upon a foundation. This was not the way our Lord Jesus Christ and His apostles understood and used the early chapters of Genesis.

In contrast to theistic evolutionism, which presumably is quite widespread among both Protestant and Roman Catholic scientists, the most rational way to explain the origin of our vastly complex solar system is in terms of a direct creation by God. And if this be a reasonable position within the revealed frame of reference of biblical theism and in view of the conspicuous failures of evolutionary alternatives, may not the supernatural origin of the astronomic system we know the best serve as a model for the supernatural origin of the stellar systems that lie beyond our own?

In other words, if God created *ex nihilo* the two great lights that rule day and night, He could also have created *ex nihilo* "the stars also" (Gen. 1:16). In the words of Paul A. Zimmerman: "The biblical account of creation by Almighty God has not been disproved by science. It remains today, even from the viewpoint of reason, I believe, the most logical, believable account of the beginning of the earth and the rest of the universe."[20]

Carl F. H. Henry is a prominent American theologian who has carefully watched the liberalizing trends within evangelical institutions of higher learning, such as those associated with the Christian College Consortium. With regard to the creation/evolution issue, he says,

> An appraisal by Albert J. Smith shows that no well-defined institutional stand exists even among schools in the Christian College Consortium

19. Howard J. Van Till, *The Fourth Day: What the Bible and the Heavens Are Telling Us About the Creation* (Grand Rapids: Wm. B. Eerdmans Pub. Co., 1986), pp. 83–85.

20 "Some Observations on Current Cosmological Theories," *Concordia Theological Monthly* 24:7 (July, 1953), p. 513.

> other than to emphasize man's special origin and dignity and God as Creator. . . . ("Creationism and Evolution as Viewed in Consortium Colleges" [*Universitatus* 2, no. 1 (March, 1974), pp. 5ff.] Smith's conclusion is that alongside their desire to avoid dogmatism most evangelical colleges interpret "the belief that all truth is God's truth" to imply that "science is a valid source of revelation. . . . Instructors insist on a personal God who is creator and sustainer in a manner largely unknown but somewhat revealed in nature through science. . . ." The verdict seems unavoidable that such a stand or nonstand can hardly hope to articulate a Christian world-life view that significantly illumines the current debate over evolutionary theory.[21]

Dr. Henry further observes,

> It is significant that the public challenge to secular humanism did not come from within the traditional evangelical colleges, many of them given to rather ambiguous and tolerant correlation of evolution with some basic theistic fundamentals. Nor did it stem from within the American Scientific Affiliation, most of whose members are theistic evolutionists. . . . It was primarily evangelical scientific creationists, who insist that creation be taken more seriously than merely as an effort to correlate Genesis and evolution, who turned public attention on the classroom inculcation of evolution as fact. . . . The fact is that a number of secular scientists are now asking more pointed questions about Darwinian evolution than are the faculties in some church-related colleges.[22]

The Purpose of the Stellar Creation

Why did God create the sun, moon, and stars on the fourth day rather than the first day? One possible explanation is that in this way God has emphasized the supreme importance of the earth among all astronomical bodies in the universe. In spite of its comparative smallness of size, even among the nine planets, to say nothing of the stars themselves, it is nonetheless absolutely unique in God's eternal purposes.

It was on this planet that God placed man, created in His image, to exercise dominion and to worship Him. It was to this planet that God came in the person of His Son nineteen hundred years ago to become a permanent member of the human race and to die for human sins upon a

21. Carl F. H. Henry, *God, Revelation and Authority*, Vol. 6, p. 149.
22. Ibid., p. 151.

rugged cross. And it will be to this same planet that this great God and Saviour will return again to establish His kingdom. Because of its positional superiority in the spiritual order of things, therefore, the earth was formed first, and then the stellar systems.

Another possible reason for this order of events is that God, by this means, made it clear that the earth and life upon it do not owe their existence to the greater light that rules the day, but rather to God Himself. In other words, God was perfectly able to create and take care of the earth and even living things upon it without the help of the sun. Apart from the Scriptures, of course, this would hardly be an obvious fact to mankind.

In ancient times (and even in some parts of the world today) great nations actually worshiped the sun as a god. In Egypt he was called *Ra*, and in Babylon he was known as *Shamash*. After all, such worship seemed quite reasonable in view of the fact that the sun provided light, warmth, and apparently, life itself.

Even the Jews were greatly tempted to enter into such worship, as may be judged by such passages as Deuteronomy 4:19 and 17:3. Job confessed:

> If I have looked at the sun when it shone, or the moon going in splendor, and my heart became secretly enticed, and my hand threw a kiss from my mouth, that too would have been an iniquity calling for judgment, for I would have denied God above (Job 31:26–28).

Perhaps it is not inappropriate to suggest that evolutionary theory provides a modern and subtle counterpart to the ancient sun-worship cult, for if we must trace our origin to the sun or to a proto-sun, and if we live, move, and have our being exclusively through its boundless blessings and provisions, *then it is our God*!

The creation account in Genesis completely undermines all such blasphemies by putting the sun in a secondary position in reference to the earth. It is not only a mere creature of God, but also a servant to man, the crown of God's creation.

But if the sun, moon, and stars are not ultimately essential to the earth's existence, then why did God create them? Three basic reasons are listed in Genesis 1:14. They are for lights, for seasons (a clock-calendar), and for signs.

As *lights*, they replaced the special and temporary light of the first three days.

As a *calendar*, dividing seasons, days, and years, they enable men to plan their work accurately into the distant future, thus reflecting the purposive mind of God.

As *signs*, they teach and ever remind men of vastly important spiritual truths concerning the Creator.

David learned from the heavens the transcendence of God and his own comparative nothingness: "When I consider Thy heavens, the work of Thy fingers, the moon and the stars, which Thou hast ordained; what is man that Thou dost take thought of him?" (Ps. 8:3). The apostle Paul insisted that men are utterly without excuse for their idolatries, for "what has been made" gives clear testimony to the "eternal power and divine nature" of the Creator (Rom. 1:20).

Apparently, the sun, moon, and stars more effectively accomplish these purposes than one great light source could have. There need be no other reason for their existence than this threefold ministry to man.

But would this not have been an unnecessary waste of God's creative energies? Isaiah gives the effective answer: "Do you not know? Have you not heard? The Everlasting God, the LORD, the Creator of the ends of the earth, does not become weary or tired. His understanding is inscrutable" (Isa. 40:28).

The heavens are the work of God's "fingers" (Ps. 8:3), and when they have fulfilled their God-intended purpose, they will flee away from His face and no place will be found for them (Rev. 20:11). The eternal city will have "no need of the sun or of the moon, to shine upon it, for the glory of God has illumined it," and the Lord Jesus Christ will be the lamp thereof (Rev. 21:23; cf. 22:5).

Christ and His Word, therefore, must be our final guide as we seek to understand the origin, meaning, and destiny of the heavens and the earth.

Conclusion

God's written revelation concerning the creation of the earth, the moon, the other planets, the sun and the stars gives special emphasis to the concept of *ex nihilo*—"out of nothing" creation. These gigantic and marvelous physical entities, in all their unending variety and beauty, circling through the vastness of space, were designed to tell us something of our God that we could not otherwise know.

Four thousand years ago God asked Job,

> "Where were you when I laid the foundation of the earth! Tell Me, if you have understanding. . . . Can you bind the chains of the Pleiades, or loose

> the cords of Orion? Can you lead forth a constellation in its season, and guide the Bear with her satellites? Do you know the ordinances of the heavens, or fix their rule over the earth?" (Job 38:4, 31–33).

In astounding contrast to the characteristically proud and secular twentieth-century mind, Job answered the Lord:

> "I know that Thou canst do all things, and that no purpose of Thine can be thwarted. . . . Therefore I have declared that which I did not understand, things too wonderful for me, which I did not know. . . . I have heard of Thee by the hearing of the ear; but now my eye sees Thee; therefore I retract, and I repent in dust and ashes" (Job 42:2–6).

Our ultimate choice is either to believe that the universe is the product of random and meaningless chance, or that it was created by a personal living God. But these alternative faith commitments cannot be equal options for men who bear the image of God indelibly imprinted upon their innermost being. The God of creation simply will not allow Himself to be compared with any other "deity," including evolutionary time/chance:

> "To whom then will you liken Me that I should be his equal?" says the Holy One. Lift up your eyes on high and see who has created these stars, the one who leads forth their host by number. He calls them all by name; because of the greatness of His might and the strength of His power not one of them is missing . . . "Turn to Me, and be saved, all the ends of the earth; for I am God, and there is no other" (Isa. 40:25–26; 45:22).

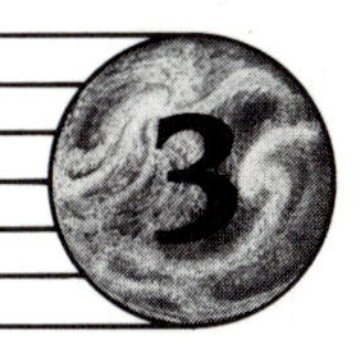

The Creation of Plants and Animals

Genesis and the Geologic Timetable

The order of events in the appearance of living things as recorded in the Book of Genesis differs profoundly from that which is generally taught today. Although some writers have sought to draw parallels between the days of Genesis and the various periods of the geologic timetable, it has become increasingly evident that the effort has been unsuccessful. A glance at the Genesis record will show why.

In the first place, Genesis puts the creation of all basic types of land plants (including *fruit trees*) in the *third* day, two days *before* the creation of marine creatures, whereas evolutionary geologists insist that marine creatures came into existence hundreds of millions of years before fruit trees. Secondly, Genesis tells us that God made the sun, moon, and stars (*made* being clearly synonymous with *created* in this chapter) on the *fourth* day, after the creation of plants, whereas evolutionists assume that the sun existed before the earth itself was formed.[1] Thirdly, Genesis states

1. Frequently it has been claimed that the sun and moon were not created on the fourth day because the Hebrew verb used in Genesis 1:16 is *'āsâh* ("made") rather than *bārā'* ("created") as in Genesis 1:1. However, this is a significant error. In creation contexts the two verbs are used synonymously, as any concordance will show. For example, marine creatures were "created" (v. 21) while land animals were "made" (v. 25). Surely this cannot mean that land animals were not created! Furthermore, these two verbs are used alternately to

that the birds were created on the *fifth* day with the fishes, but the evolutionary timetable has birds following reptiles (which were not created until the sixth day). And finally, Genesis puts the creation of insects ("creeping things") in the *sixth* day, three days after flowering plants were created; but this would be impossible if the days were ages, for some pollination requires insects. After carefully analyzing the entire question, one respected theologian/scientist concluded: "There are far more differences between the Genesis account and the [evolutionary interpretation of the] geological record than there are similarities. And these differences are quite significant. . . ."[2]

Because of such discrepancies, some who have sought a parallel or correspondence between the order of events in Genesis 1 and the order of events in the timetable of evolutionary geology ("concordism") are now suggesting that the "days" of Genesis 1 are not really consecutive after all. Robert C. Newman, for example, feels that "the 'days' of Genesis 1 were 24-hour days, sequential but not consecutive, and that the creative activity largely occurred *between* days rather than *during* them. That is, each Genesis day introduced a new creative period;" with each period lasting many millions of years and the seventh "day" still future![3]

Davis A. Young, on the other hand, sees the "days" of Genesis 1 as "seven successive figurative days of indeterminate duration." Furthermore, Young goes so far as to allow the events in these "days" to overlap

describe the very same events: Genesis 1:26 ("make") and 1:27 ("created"); Genesis 2:4a ("created") and 2:4b ("made"); Genesis 1:1 ("created") and Exodus 20:11 ("made"); Genesis 1:16 ("made") and Psalm 148: 3, 5 with Isaiah 40:26 ("created"). After a careful study of these verbs, Bruce K. Waltke concluded: "It is clear that *'āsâh* and other verbs may designate creation by fiat *ex nihilo*. The doctrine of *creatio ex nihilo* does not depend on the verb *bārā'*. . . . The sun, moon and stars came into existence at the sole bidding of their Creator" ("The Creation Account in Genesis 1:1–3" [*Bibliotheca Sacra* 132:528 (October, 1975)], 337). Thus, it is exegetically erroneous to date the creation of the sun and moon before the fourth day of creation week. Cf. Weston W. Fields, *Unformed and Unfilled: A Critique of the Gap Theory* (Phillipsburg, NJ: Presbyterian and Reformed Pub. Co., 1976), pp. 53–74; and Whitcomb and DeYoung, *The Moon: Its Creation, Form and Significance*, p. 72. See below, chapter 5, "The Gap Theory."

2. John W. Klotz, *Modern Science in the Christian Life* (St. Louis: Concordia Publishing House, 1961), pp. 111–12. Carl F. H. Henry lists twenty-nine major differences (*God, Revelation and Authority*, 6:147–48.

3. Newman and Eckelmann, *Genesis One and the Origin of the Earth*, pp. 74, 65.

each other. Thus, we now have not only a "modified intermittent-day model" but also an "overlapping day-age model."[4]

Such imaginative revisions of the once popular concordist concept simply demonstrate, in the opinion of many, the basic artificiality of the entire Day-age Theory. In the words of one theistic evolutionist, a professor (and colleague of Davis Young) at Calvin College,

> The concordist interpretation is another example of the meaningless explanations that derive from addressing inappropriate questions to Scripture. It is the sterile offspring of the unwarranted assumption that Genesis 1 is meant to be a journalistic recounting of God's original creative activity.[5]

Like the old Ptolemaic astronomy, concordism has been killed by a thousand qualifications.[6] But such a fate was inevitable, for concordism, like theistic evolution, ignores the supremely important truth that physical death entered the world only after Adam sinned (cf. Rom. 5:12; 8:18–23).[7] Thus, it is obvious that God never intended Genesis to harmonize with evolutionary concepts of earth history.

Many theologians, with weak commitments to the authority of Scripture, and sensing the failure of the various concordist theories to actually harmonize Genesis with evolutionary geology, have drifted into various forms of theistic evolutionism such as the neo-orthodox approach,

4. Davis A. Young, *Creation and the Flood: An Alternative to Flood Geology and Theistic Evolution*, pp. 89, 116–17. Pattle P.-T. Pun, a biologist at Wheaton College, has adopted *both* of these concordist models (*Evolution: Nature and Scripture in Conflict?* [Grand Rapids: Zondervan Publishing House, 1982], pp. 261–66).

5. Howard J. Van Till, *The Fourth Day*, p. 91. Henri Blocher, also a theistic evolutionist, concludes that "the concordist boat is wrecked" on the reef of the fourth day of creation, which places the creation of the Sun after the creation of trees (*In the Beginning* [Downers Grove, IL: InterVarsity Press, 1984], p. 45).

6. Richard H. Bube, a prominent theistic evolutionist, claimed that the only consistent alternative to the "instantaneous creation of a young earth" is total evolutionism under the providential direction of God. On this basis, he strongly criticised the efforts of Davis Young, Robert Newman, and Herman Eckelmann to settle at "a way station on the progression of thought" from strict creationism to theistic evolutionism (*Journal of the American Scientific Affiliation* 30:2 [June, 1978], p. 91).

7. The problem of sin and death is the main focus of Fred Van Dyke's critique: "Theological Problems of Theistic Evolution" (*Journal of the American Scientific Affiliation* 38:1 [March, 1986], pp. 11–18). See also E. H. Andrews, *Christ and the Cosmos* (Evangelical Press, 16 High St., Welwyn, Herts, AL6 9EQ England), pp. 86–90.

which lifts the entire creation account out of the realm of history and science, and denies that God ever intended these words to convey anything other than "theological" perspectives.

Bernard Ramm is one of the best-known representatives of this school of thought, which reduces Genesis 1–2 to the level of a mere "literary framework" for communicating "ultimate truth." He has now abandoned the effort to find either valid history or dependable science in the opening chapters of the Bible. Seeing the collapse of concordism (the view he formerly espoused[8]) and disdaining any return to traditional views of a recently created earth, he concludes that we must learn to radically rethink creation in the categories of neo-orthodoxy as expressed in the writings of Herrmann, Giersch, and Barth. Dr. Ramm thinks that

> . . . a truly Biblical and truly theological notion of creation is going to come from these circles, and not from the surreptitious notion in American orthodox and fundamentalist circles that Genesis I is only revelation or inspired if it in some way anticipates modern science.[9]

In refutation of the framework hypothesis of Genesis 1, it may be stated, in the first place, that it finds no place in true biblical hermeneutics. Edward J. Young, in his important work entitled, *Studies in Genesis One,*[10] insisted that the early chapters of Genesis bear none of the marks of poetry or saga or myth, but must be interpreted as literally as any other "straightforward, trustworthy history" recorded in Scripture.[11] Furthermore, he demonstrated that the biblical text calls for a chronological succession of distinct time periods.[12] We have elsewhere pointed out why these time periods must have lasted approximately twenty-four hours. Thus, God never intended to parallel the days of creation with the so-called ages of historical geology.

This leads us to our second basic objection to the framework hypothesis. Like the Day-age Theory which it has increasingly replaced, it cannot take seriously the perfection of God's completed creation as

8. Cf. *The Christian View of Science and Scripture* (Grand Rapids: Wm. B. Eerdmans Pub. Co., 1954).

9. "Comments on Mr. Verduin's Essay," *Christianity Today* (May 21, 1965), p. 15.

10. *Studies in Genesis One* (Phillipsburg, N.J.: Presbyterian and Reformed Pub. Co., 1964).

11. Ibid., p. 105.

12. Ibid., pp. 77–100.

stated in Genesis 1:31—"God saw *all* that He had made, and behold, it was *very good*." How can we believe, in the light of this statement, that many kinds of plants and animals had already become extinct during the billion or more years of "struggle for existence" that supposedly preceded Adam's creation? Was not Adam told to exercise dominion over "every living thing" (Gen. 1:28) in a unique sense that is no longer true today (cf. Heb. 2:8)? Did not Adam find himself in a world of exclusively *herbivorous* animals (Gen. 1:30; cf. Isa. 11:7)? But the framework theory presupposes that many animals in Adam's day were carnivorous and had been so for hundreds of millions of years. How can this be reconciled with the fact that "the creation was subjected to futility" and "groans and suffers the pains of childbirth together until now" and is in "slavery to corruption" as a result of the Edenic Curse upon the animal kingdom *following Adam's fall* (Rom. 5:12; 8:20–22)?

Bernard Ramm, among others, has sought to solve the problem by redefining the word "good" in Genesis 1:31:

> The universe must contain all possible ranges of goodness. One of these grades of goodness is that it can fail in goodness . . . If there were nothing corruptible, or if there were no evil men, many good things would be missing in the universe. The lion lives because he can kill the ass and eat it. . . . The entire system of nature involves tigers and lions, storms and high tides, diseases and parasites.[13]

In other words, since it is difficult to imagine any other balance of nature than the one which we observe today, Ramm assumed that the world has always been this way.

G. C. Berkouwer's description of the "harmonistic theodicy" of the Stoics and of the German philosopher Leibnitz (1646–1716) accurately fits this world view. Berkouwer observes:

> This theodicy [a vindication of the justice of God in permitting evil to exist] rests principally on a relativizing of sin. God's goodness shines only as the grim clouds of sin and evil are dispelled. . . . Recall, in contrast, how the Scriptures speak of sin as having "entered into the world" (Rom. 5:12), as "enmity against God" (Rom. 8:7). The basic error of this theodicy is its fundamental assumption that reason can find a proper place for sin in creation . . . a fundamental failure to appreciate the awful reality of sin,

13. *The Christian View of Science and Scripture* pp. 93–95.

"The Mountains Rose"
(Glacier National Park, Montana)

It was not gradually, throughout long ages before the creation of mankind, but suddenly, at the end of the Genesis Flood, that the continents and "the mountains rose" (Ps. 104:8). The context of this statement, namely, the clear reference to the rainbow covenant of Genesis 9 in the next verse, makes it obvious that the author is here referring to the gigantic continental uplifts that occurred *just after the Flood* (cf. Gen. 8:1–3). Some have proposed that Psalm 104 is referring to events during the third day of creation week, but; as David G. Barker has demonstrated,

> In the light of the broader cosmological perspective of the psalm and the similar citation in Isaiah 54:9, verses 6–9 [in Psalm 104] clearly point to the Noahic deluge of Genesis 6–9 rather than the creation account of Genesis 1. To relegate these verses to the creation account creates serious theological and historical problems, especially in the light of the emphatic statements regarding the finality of the determination of the oceanic boundaries. . . . It was the mountains that went up and the valleys [basins] that went down. This provides valuable insight into the catastrophic tectonic activities of the flood year" ("The Waters of the Earth: An Exegetical Study of Psalm 104:1–9" (*Grace Theological Journal* 7:1 [Spring, 1986], p. 80).

If the great mountain ranges of the present world (the Rockies, the Andes, the Alps, the Himalayas) rose to their enormous heights suddenly, just after the Flood, men have all the more reason to glorify the power of Almighty God, "For the LORD is a great God, and a great King . . . the peaks of the mountains are His also. . . . And His hands formed the dry land" (Ps. 95:3–5; cf. Ps. 65:5–8; Amos 4:13).

Mountain-building is one of the unsolved mysteries of modern uniformitarian geology; but the Bible supplies the missing dynamic in terms of God's omnipotent intervention at the closing phase of the Flood year. Before the huge sedimentary deposits laid down during the Flood had time to consolidate or solidify they were pushed up to great heights. Still plastic in consistency, they did not split or shatter when uplifted, but rather were bent and twisted like pages in a thick magazine. This illustrates the important fact that some major geologic features simply cannot be explained in terms of gradual processes over long ages. Natural revelation through science desperately needs the guidelines of special revelation through Scripture if genuine answers to nature's mysteries are ever to be found.

> suffering, and death. Oversimplification typifies it, and the self-evidency of this oversimplification has contributed to modern man's profound distrust of every attempt at a theodicy.[14]

A third major defect in the framework hypothesis (as well as the concordist or Day-age Theory) is that its advocates assume the validity of the timetable of uniformitarian geology. This scheme of earth history was devised in the early nineteenth century by men who rejected the biblical testimony to a universal Flood, and who sought to explain the earth's features in terms of the gradual geologic processes that were observable in their day.[15] It is obvious, however, that the uniformity principle is completely inadequate for interpreting fluviatile plains, enclosed lake basins, raised river terraces, incised meanders, mountain building, vast horizontal superimposed beds of fossil plants and animals, huge lava plateaus, continental ice sheets, frozen mammoths, and the great reversed-order sequences ("overthrusts") of Montana, Wyoming, and Switzerland. The geologic timetable involves circular reasoning, for it assumes the truth of total organic evolution to arrive at the dates assigned to index fossils and the rocks that contain them. It hardly seems necessary, therefore, to mold Genesis into conformity with a scheme that has failed both logically and experimentally.[16]

The Original Abundance of Life

To those who have been taught to believe that life in the oceans began with a single-celled creature, the Genesis account presents an astounding picture: "Then God said, 'Let the waters teem with swarms of living creatures. . . . And God created the great sea monsters, and every living creature that moves, with which the waters swarmed, after their kind . . ." (Gen. 1:20–21).

It may be noted that in the enumeration of sea creatures in Genesis 1:21 "great sea monsters" ("great whales," KJV) are mentioned first. It is

14. *The Providence of God* (Grand Rapids: Wm. B. Eerdmans Pub. Co., 1952), pp. 238 f.

15. Cf. R. T. Clark and J. D. Bales, *Why Scientists Accept Evolution* (Phillipsburg: N.J.: Presbyterian and Reformed Pub. Co., 1966), p. 19.

16. Cf. John C. Whitcomb, Jr. and H. M. Morris, *The Genesis Flood* pp. 116–211; N. A. Rupke, "Prolegomena to a Study of Cataclysmal Sedimentation," in *Why Not Creation?* ed. Walter Lammerts (Phillipsburg, N.J.: Presbyterian and Reformed Pub. Co., 1970), pp. 141 ff.; and Joseph C. Dillow, *The Waters Above: Earth's Pre-Flood Vapor Panopy*, pp. 311–420.

just as if God were intentionally contrasting His great powers of creation with the evolutionary hypothesis which He knew would eventually dominate the world of scientific speculation concerning origins. The blue whale is the largest animal that has ever lived, some individuals attaining a length of 110 feet and a weight of 300,000 pounds. For the God of creation, this was no problem at all, for He created the earth out of nothing by a mere word.

The theory of evolution is deeply embarrassed by the existence of aquatic mammals such as whales, for it must assume that these monsters of the deep evolved from four-legged pig-like land mammals which in turn had evolved from reptiles and fishes. This assumption is not only completely lacking in genetic and paleontologic evidence but is logically absurd. Not only so, but the failure of evolutionary theory to account for even the first speck of life has become increasingly evident with the passing years.

Several years ago a symposium of papers by some of the world's leading experts on the origin of life was published under the title *The Origins of Prebiological Systems.*[17] One of the papers, presented by Peter T. Mora of the Macromolecular Chemistry Section of the National Institute of Health in Bethesda, Maryland, entitled, "The Folly of Probability," caused considerable debate among the scientists present at the meeting, because he showed that probability statistics offer no hope in explaining the origin of a unicellular organism from inorganic chemicals:

> I believe we developed the practice of infinite escape clauses to avoid facing the conclusion that *the probability of a self-reproducing state is zero*. This is what we must conclude from classical quantum mechanical principles, as Wigner demonstrated (1961). These escape clauses postulate an almost infinite amount of time and an almost infinite amount of materials (monomers), so that even the most unlikely event could have happened. This is to invoke probability and statistical considerations *when such considerations are meaningless*. When for practical purposes the condition of infinite time and matter has to be involved, *the concept of probability is annulled*. By such logic we can prove anything, such as that no matter how complex, everything will repeat itself, exactly and innumerably.[18]

17. *The Origins of Prebiological Systems*, ed. S. W. Fox (New York: Academic Press, 1965).
18. Ibid., p. 45.

Whales

Whales, along with all marine creatures, appeared on the fifth day of creation, and thus preceded the land mammals from which they supposedly evolved. The God of creation had no problem launching these monsters of the deep without the help of vast periods of time or previously existing similar forms. But evolutionists have enormous problems explaining how such complex and uniquely structured animals could have evolved into their present forms. To list only three examples out of many: (1) "The female whale gives birth to her young under water, and suckles them under water. The baby whale has its windpipe prolonged above the gullet to prevent milk ejected out of its mother's mammary glands from getting into its lungs. Further, the baby whale's snout is cunningly arranged to fit a receptacle on the body of its mother into which she secretes milk. In this manner the baby whale is prevented from imbibing sea-water with its mother's milk." Could such organs have evolved by random mutations and natural selection? (2) The whale's eye "differs from that of land mammals in having the eyeball immovable, eyelids without eyelashes, no tarsus in the eyelid, a downward direction of eye axis, a more spherical lens, and a greatly thickened sclera." (3) The ear of a whale "is clearly constructed on a different plan from that of the mammalian ear for the reception of air-borne sound waves. The cetacean ear operates in water and is able to resist temporary high pressures when the animal is at depth" (Quotes from Frank Cousins, *Evolution Protest Movement,* No. 114, April, 1964). Evolutionists often point to vestigial hind legs near the pelvis. But these are found *only* in the Right Whale, and upon closer inspection turn out to be strengthening bones to the genital wall. Michael Pitman says, "Evidence for the evolution of either toothed or baleen whales is absent" (*Adam and Evolution* [1984], p. 211. See also Pitman's nine arguments in support of the direct creation of whales (pp. 212–13). Thus we appreciate more than ever the psalmist's challenge: "Praise the LORD from the earth, sea monsters and all deeps" (Ps. 148:7).

In a review of this symposium entitled, "Did Life Evolve?" R. L. V. Ulbricht struck hard at the notion that "the origin of life from non-life can be explained by invoking time and chance."[19] He concluded by stating:

> The reviewer willingly admits that it is easier to criticize than it is to suggest better alternatives. However, the pace of development in molecular biology is such that in the forseeable future the time will come when the ability of science to solve this problem will become more and more crucial, and failure might mean the beginning of a new revolution in thought.[20]

We might add here that the time for this revolution has already come, for a hundred years of searching for answers to the origin-of-life problem have ended in complete failure. Multiplying zero creative power by five billion or even an infinite number of years still equals zero results.

The Nobel Prize winner Francis Crick, though an atheist, attained a "moment of truth" when he confessed:

> An honest man, armed with all the knowledge available to us now, could only state that in some sense, the origin of life appears at the moment to be almost a miracle, so many are the conditions which would have had to have been satisfied to get it going.[21]

One can begin to understand the frustrations of such scientists when it is realized that "a simple one-celled bacterium, R. *coli*, contains DNA information units that are the equivalent of 100 million pages of *Encyclopedia Britannica*![22]

In an important work by Malcolm Dixon and Edwin Webb entitled *Enzymes* the authors demonstrated that in view of the fact that enzymes

19. "Did Life Evolve?" *Chemistry and Industry* (January 8, 1966), p. 44.

20. Ibid., p. 45.

21. *Life Itself* (New York: Simon and Schuster, 1981), p. 88; quoted in Michael Pitman, *Adam and Evolution*, p. 268. See also C. B. Thaxton, et al, *The Mystery of Life's Origin: Reassessing Current Theories* (New York: Philosophical Library, 1984); S. E. Aw, *Chemical Evolution: An Examination of Current Ideas* (El Cajon, CA: Master Books, 1982); and A. E. Wilder-Smith, *The Creation of Life: A Cybernetic Approach to Evolution* (Wheaton, IL: Harold Shaw Publishers, 1970); and *The Natural Sciences Know Nothing of Evolution* (San Diego: Master Books, 1981).

22. Carl F. H. Henry, *God, Revelation and Authority*, Vol 6, p. 177, citing R. L. Wysong, *The Creation-Evolution Controversy* (Inquiry Press, P.O. Box 1766, East Lansing, MI 48823), pp. 114ff.

can only be formed by other enzymes, there is no known way for life to have started in the first place.[23] After enumerating some of the insurmountable problems in the evolutionary concept, the authors concluded:

> A further difficulty is that of holding the components of the system together until a cell membrane is formed, assuming life to have begun in the ocean. Unless the ocean contained throughout a fairly high concentration of the components (thus being itself one gigantic living cell!), the components would rapidly disperse, as happens now when a cell membrane is ruptured. The system would then perish "by lethal dilution." But the formation of a cell membrane implies a system which already had a high degree of organization. Thus the whole subject of the origin of enzymes, like that of the origin of life, which is essentially the same thing, bristles with difficulties. We may surely say of the advent of enzymes, as Hopkins said of the advent of life, that it was the most improbable and the most significant event in the history of the Universe.[24]

Thus, Pasteur's demonstration that life can only come from life stands stronger than ever, and the creation account of Genesis is thereby confirmed.

Michael Pitman explains,

> Enzyme systems are doing every minute what battalions of fulltime chemists cannot. . . . Can anyone seriously imagine that naturally occurring enzymes realized themselves, along with hundreds of specific friends, by chance? Enzymes and enzyme systems, like the genetic mechanisms whence they originate, are masterpieces of sophistication. Further research reveals ever finer details of design. . . . Dixon confesses that he cannot see how such a system could ever have originated spontaneously."[25]

23. *Enzymes*, 2nd ed. (New York: Academic Press, 1964), p. 665.

24. Ibid., p. 669.

25. *Adam and Evolution* (Grand Rapids: Baker Book House, 1986) pp. 144f. His reference to Dixon may be found in Dixon, Tipton, Thorne and Webb, *Enzymes*, 3rd ed. (New York: Longman, Green and Co., 1979), pp. 656ff.

The Experiments of Louis Pasteur

It was my privilege to visit the museum of the Pasteur Institute in Paris in 1975 and to see the flasks pictured here. It was by means of two years of experiments carried out in this laboratory and in these very flasks that Louis Pasteur demonstrated in 1861 the impossibility of spontaneous generation of life, a view promoted by Charles Darwin only two years earlier in his book, *The Origin of Species*. Pasteur totally rejected Darwin's views, because he believed that "to bring about spontaneous generation would be to create a germ. It would be creating life . . . God as author of life would then no longer be needed. Matter would replace Him. God would need to be invoked only as author of the motions of the universe" (Quoted in Ian T. Taylor, *In The Minds of Men* [1984], p. 182). "Pasteur summarized his work with a triumphant lecture at the Sorbonne [in Paris] in 1864, which he concluded with this remark: 'Never will the doctrine of spontaneous generation recover from the mortal blow of this simple experiment" (Robert Shapiro, *Origins: A Skeptic's Guide to the Creation of Life on Earth* [New York: Summit Books, 1986], p. 52). Louis Pasteur's prediction was correct. The evolutionary theory that life originated by chance is dead.

A Monkey and a Typewriter

How long would it take lifeless chemicals under ideal conditions to evolve into a living protozoan? Answer: it could *never* happen! Let's simplify the problem. How long would it take a monkey, pounding irrationally at a typewriter, to come up with the words of Genesis 1:1 ("In the beginning God created the heavens and the earth")? In fact, let's allow a million tireless monkeys to pound away at record speed (twelve keys a second) on simplified typewriters with only capital letters.

> Try to think of a rock so large that if the earth were at its center its surface would touch the nearest star. This star is so far away that light from it takes more than four years to get here, traveling 186,000 miles every second. If a bird came once every million years and removed an amount equivalent to the finest grain of sand, *four such rocks* would be worn away before the champion super simians would be expected to type Genesis 1:1 (Bolton Davidheiser, *Evolution and Christian Faith,* Presbyterian and Reformed Publishing Co., 1969, p. 363; using the calculations of William Feller, *An Introduction to Probability Theory and Its Implications,* Wiley, 1950, I, 266).

When the absurdities of evolutionary improbabilities are fed into modern computers, red lights flash and the machinery jams! See Paul S. Moorehead, editor, *Mathematical Challenges to the Neo-Darwinian Interpretation of Evolution* (The Wistar Institute Press, 1967). Sir Fred Hoyle stated that the chance of life evolving from non-living matter is comparable with the chance that "a tornado sweeping through a junk-yard might assemble a Boeing 747 from the materials therein" (*Nature* Vol. 294 [Nov. 12, 1981], p. 105).

To be sure, vast publicity has accompanied Stanley Miller's experiment in forming amino acids in an apparatus containing methane, ammonia, hydrogen and water and energized by an electric discharge, as evidence that life could have evolved out of inorganic chemicals in the ancient oceans. However, Duane T. Gish, a former research biochemist at Upjohn Laboratores, observed, "The significance of this demonstration is not really very great at all; it might even be termed trivial. Having placed a selected number of gasses in a closed system and supplied a source of energy we would rather be surprised had *not* such a variety of carbon, oxygen, and nitrogen-containing compounds been formed."[26] Dr. Gish then referred to a paper by Philip Abelson, Director of the Geophysical Laboratory, Carnegie Institution of Washington, to the effect that such a reducing atmosphere would have been thermodynamically impossible because, "An analysis of geologic evidence sharply limits the areas of permissible speculation concerning the nature of the primitive atmosphere and ocean."[27] Gish concluded: "It is evident, then, that the basis for Miller's experiment did not exist."

Thus, whereas the God of creation was able to create whales and "every living creature that moves" in a moment of time without exhausting His energies at all, evolutionists cannot imagine how a single-celled self-replicating organism could have been generated spontaneously, even with the kind of reducing atmosphere they have prepared in laboratories and with infinite time theoretically at their disposal. Can a greater contrast between two world views be imagined?

To state this contrast in different terms, Genesis tells us that *all the basic kinds* of creatures that have ever lived appeared almost simultaneously at the beginning of earth history, and that there have been fewer and fewer *kinds* (not varieties) ever since, as many have become extinct through the struggle for existence in a world that groans under the bondage of corruption (Gen. 2:1–3; cf. Rom. 8:20–22). On the other hand, the theory of evolution requires one solitary, submicroscopic speck of life at the beginning, with more and more kinds of organisms emerging as the ages come and go. By faith, the evolutionist understands that the organic world was framed by chance in contradiction to present biologic proc-

26. "Critique of Biochemical Evolution," in *Why Not Creation?* ed. Walter Lammerts, p. 284.

27. *Abstracts, 133rd National Meeting*, American Chemical Society, (April 1958), p. 53. Cf. Michael Pitman, *Adam and Evolution*, pp. 138–39.

esses. By faith, the Christian understands that the organic world was created by the spoken word of an infinitely powerful, omniscient, personal God as revealed in the infallible Scriptures and as seen in the obvious discontinuities in the fossil record and in biologic systems today.[28]

The Limits of Variation

It is a foundational principle of the theory of evolution that there can be no fixed limits to the possibility of variation in living things, for the theory assumes that all living things in the world today, both plants and animals, have developed from a single-celled organism. This is the family tree concept of living things, which confronts the student in most textbooks that deal with life sciences, historical sciences, and even world history. There are no major institutions of higher learning anywhere in the world today (as far as I am aware) that offer advanced degrees in the natural sciences where the family tree concept of general evolution is rejected. And yet amazingly enough, a century of research on the part of thousands of specialists has failed to produce any clear evidence in contradiction to the biblical doctrine that living things were created to reproduce after their kind.

Instead of a single family tree of living things, the Bible presents the picture of a great *forest of trees* of living things, each "tree" supernaturally created with the genetic potentialities for variations or branches, but within the strict confines of the created identity of the "tree." Thus mankind was created with potentialities for variation into many races, as distinct from each other as the nine-foot Anakim of ancient Palestine and the four-foot Pygmies of central Africa. But there has never been any serious question that men are men and that the various races belong to the same family tree.

It is apparent that God created certain kinds of animals with an even greater potentiality for variation than is true of mankind. For example, during the past few centuries as many as two hundred breeds of dogs have been developed, as different from each other as the Great Dane and

28. See the remarkably clear demonstrations of these basic discontinuities in Michael Denton, *Evolution: A Theory in Crisis* (Bethesda, MD: Adler & Adler, 1986); cf. Lane P. Lester and Raymond G. Bohlin, *The Natural Limits to Biological Change* (Grand Rapids: Zondervan Pub. House, 1984).

A Forest of Family Trees

In total contrast to the evolutionary view that all living things on this planet developed gradually (or even in spasms) through billions of years, like a gigantic tree from one speck of life, the biblical model teaches us that God directly created a vast forest of permanently separate "trees of life." According to this creationist perspective, all of the basic kinds (*baramin* = created kinds) of living things that have ever existed (such as men and gorillas and dogs and cats) were created within less than one week and have reproduced "after their kind" since then (Gen. 1; Lev. 11). God did create these "kinds" with rich potential for genetic variation into races, breeds, hybrids, etc. But so far from developing into new kinds, or even improving existing kinds, such variations are *always* characterized by intrinsic genetic weakness of individuals, in accordance with the outworking of the second law of thermodynamics through gene depletion and the accummulation of harmful mutations. Thus, the changes that occur in living things through time are always within the strict boundary lines of the created kinds, and always move toward ultimate extinction. The Genesis Flood drastically reduced, but did not destroy this amazing potential for variation.

A Forest of Family Trees
(Creationist View)

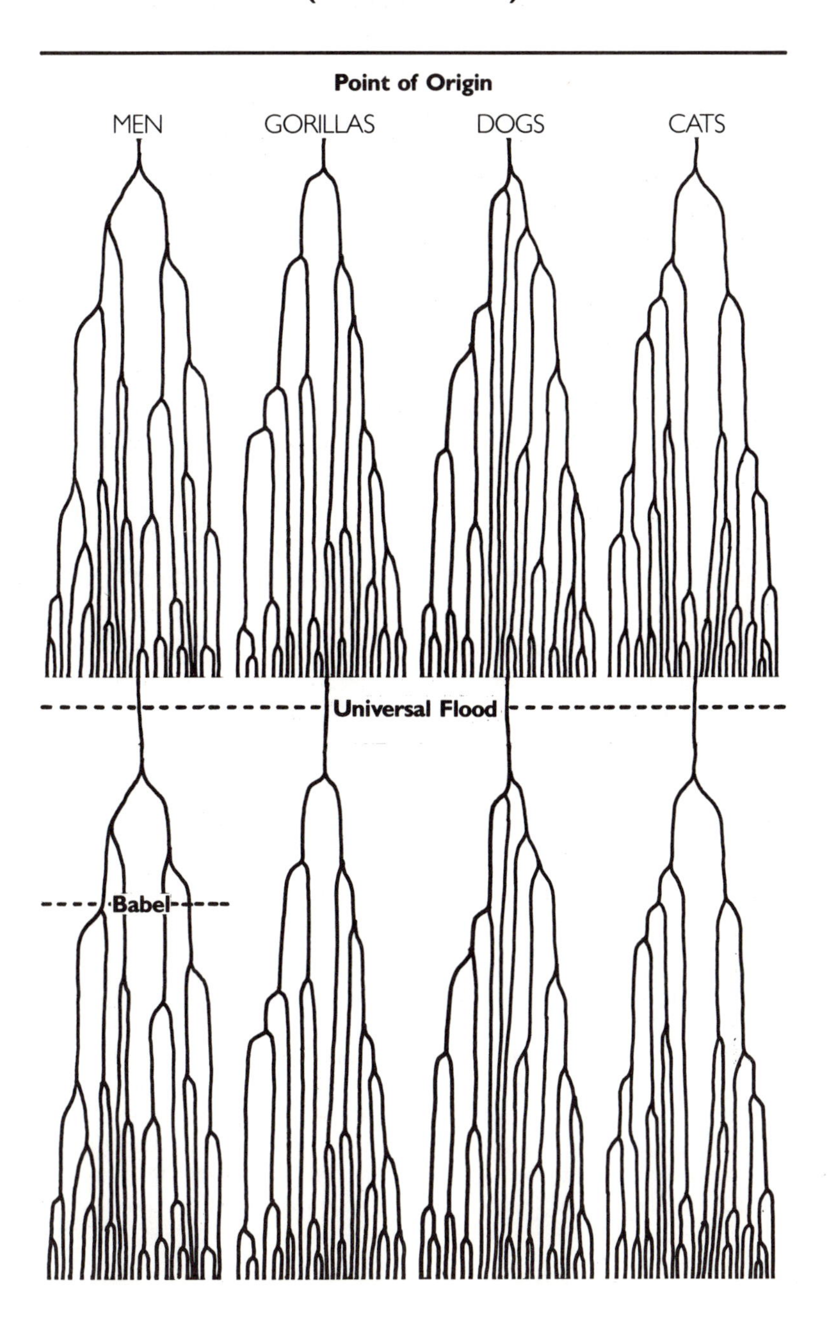

the Dachshund, but they all belong to the same created kind. This is not an evidence for evolution; it is just the opposite, for most of these variations reduce the ability of the animal to survive in nature. It is not by mutations, but rather by *recombination of extant genetic material*, that new races come into existence. As J. J. Duyvené de Wit explains,

> When a border population is pioneering for a new territory it cannot take along all the genes of its mother-population, but only part of it. Every new race or species which develops from a preceeding one, therefore, owns a *depauperized* gene pool. Hence depletion (decimation) of its gene pool, resulting from genetic drift, is the price which every race and species must pay for the privilege of coming into existence . . . the tragic fate of extremely adapted and specialized species and races, is, therefore, irrevocably: *genetic death*.[29]

Recombination thus subdivides and weakens the genotype and cannot bring about the transformation of one basic genotype into another one.[30]

Although *morphology* (structure and shape of bodies), *interfertility* (capability of a male and female mating and producing an embryo), and *chromosome count* are useful methods of identifying distinct kinds of living things, the lines in some cases are not so clearly drawn. Some evangelical scientists, in fact, have made broad concessions to evolutionary theory by suggesting that "God created the orders, and natural selection took it from there," and that "in some cases classes or even phyla could be applied.[31]

At this point, it is important to recognize that the Scriptures do provide some limits as to how inclusive the created kinds really are. In a study of the term "kind" in both Genesis 1 and Leviticus 11, J. Barton Payne concluded that "*min* must refer to subdivisions within the types of life described and not to the general quality of the types themselves."[32]

29. A *New Critique of the Transformist Principle in Evolutionary Biology* (Kampen, Neth.: Kok, 1965), p. 55.

30. Cf. Ernst Mayr, *Animal Species and Evolution* (Cambridge: Harvard University Press, 1963), p. 518; cited by Duyvené De Wit in A *New Critique of the Transformist Principle in Evolutionary Biology*, p. 54.

31. J. O. Buswell, III, "A Creationist Interpretation of Prehistoric Man," in *Evolution and Christian Thought Today*, ed. Russell Mixter (Grand Rapids: Eerdmans, 1959), p. 183.

32. "The Concept of Kinds in Scripture," *Journal of the American Scientific Affiliation* 10:2 (June 1958), p. 18. See also J. B. Payne, *The Theology of the Older Testament* (Grand Rapids: Zondervan, 1962), p. 137.

Thus, with regard to Genesis 1, "while *min* does not here require the separate creation by God of each species, it does require at least the separate creation of families within orders." With regard to Leviticus 11, however, he demonstrated that the "kinds" of birds are extended at least to genera. "Furthermore, *min* has been shown to be a term for technical enumeration; and it is used in no other, more conversational, way in Scripture. Hebrew lexicons unite in stating that *min* in Scripture has one, and only one meaning, namely 'species.'"[33]

In addition to the limitations on the term "kind" in Leviticus 11, we also have the basic guideline of the size of Noah's ark. The purpose of this structure was to save alive from a universal flood two of every "kind" of air-breathing creature (Gen. 6:19–20, 8:17). Ernst Mayr estimated that there are about 17,600 species of mammals, birds, reptiles, and amphibians in the world today. Assuming the average size of these animals to be about that of a sheep (there are only a very few really large animals, of course), this would allow room not only for two of each *species* of air-breathing animals in the world today, but also for thousands of species that have become extinct since the Flood. It seems quite obvious, therefore, that Genesis "kinds" cannot be equated with taxonomic "orders" if an ark of such magnitude had to be constructed to provide for two of each "kind."[34]

Some evangelical scientists have insisted that the evolution of the horse family (*Equidae*) provides strong proof that Genesis "kinds" were quite broad. However, G. A. Kerkut, in discussing the supposed evolution of the horse, says that "the actual story depends to a large extent upon who is telling it and when the story is being told."[35] He says,

> At present it is a matter of faith that the textbook pictures are true, or even that they are the best representations of the truth that are available to us at the present time. One thing concerning the evolution of the horse has become clear . . . instead of a family tree the branches of the tree have increased in size and complexity till the shape is now more like a bush than a tree. In some ways it looks as if the pattern of horse evolution might be even as chaotic as that proposed by Osborn (1937, 1943) for the evolution of the Proboscidea, where, "in almost no instance is any known

33. Ibid., p. 19.
34. Cf. Whitcomb and Morris, *The Genesis Flood*, p. 69.
35. *Implications of Evolution* (New York: Pergamon Press, 1960), p. 144.

Created Kinds, Noah's Ark, and Railroad Box Cars

Noah's ark was the greatest structure ever built to float upon the waters of the seas until the late nineteenth-century metal ocean-going vessels were first constructed. It was a barge, not a ship with sloping sides, and therefore had one-third more carrying capacity than a ship of similar dimensions. Assuming the minimum length of the cubit (18 inches), the ark had a capacity of nearly 1,400,000 cubic feet, and was therefore so huge that 522 modern railroad box cars could be fitted inside! And since two each of all air-breathing creatures in the world today could be comfortably carried in only 150 box cars, there was plenty of room in Noah's ark for all the kinds alive today, plus two each of extinct air-breathing types, plus food for them all.

The magnitude of the ark thus gives us important clues as to the number of air-breathing "kinds" God created in Genesis 1. The animal kingdom did not develop from a few hundred original "kinds," or from just one!

The enormous size of this flat-bottomed, square-sided barge really settles the question of whether the Book of Genesis intends to teach the concept of a universal Flood; for such a structure would not have been needed for saving animals through a localized flood situation. In fact, there would have been no need for an ark at all, for Noah's family (to say nothing of the animals) could easily have been directed by God to migrate to some region unaffected by a local flood. Since God does not give men commands that are foolish or unnecessary, we may be sure that the ark was essential for the survival of air-breathing creatures through this great, year-long catastrophe. See Whitcomb and Morris, *The Genesis Flood,* Presbyterian and Reformed Publishing Company, 1961, pp. 65–70.

CHESAPEAKE
&
OHIO

Dachshund and Great Dane

The vast physical and temperamental differences that exist among the nearly two hundred varieties of dogs (all capable of interbreeding) provide for us a perfect illustration of the richness of some God-created gene pools. Spaniels, terriers, beagles, greyhounds, bulldogs, collies, Chihuahuas, chows, and whippets—vastly different in size, shape, color pattern, hair type, and capacities, but all belonging to the same "tree" of dog-kind! Many branches, but *one tree.* God created the DNA code of this "tree" to read D-O-G, and while the world remains, no dog will begin to become a cat, nor will any cat begin to become a dog. If the present world were suddenly overwhelmed by a flood of waters, evolutionary paleontologists of a future age (if evolutionism itself has not become extinct by then!) will doubtless assume that the fossils of dachshunds must be dated a million years earlier than the fossils of Great Danes! In similar fashion, the evolution of the horse and of man have been "reconstructed." Variation within the kinds is the exact opposite of evolution, for the boundary lines established by God can *never be crossed* and the new variations that do appear (through gene recombination) represent *an essential weakening* of the individuals in these isolated varieties.

form considered to be a descendant from any other known form; every subordinate grouping is assumed to have sprung, quite separately and usually without any known intermediate stage, from hypothetical common ancestors."[36]

In the light of this, it hardly seems to be a mark of evangelical scholarship to use the supposed evolution of the horse as a basis for determining our definition of Genesis "kinds."

What are some of the limitations to variation in plants and animals that scientists have discovered in the past century? In the first place we have the laws of Mendel which are basic to the science of genetics. It has been said that Darwin would never have won the world to his position if Mendel's discoveries had received the recognition they deserved.[37] These laws explain how variations can normally occur only within fixed limits, in harmony with "after its kind" creation. In the second place, abnormal changes, or "mutations," are practically all harmful or deadly to an organism, as abundantly illustrated in the experiments upon Drosophila fruit flies. George Gaylord Simpson has written,

> If the mutation rate were .00001 (1 in 100,000—an average mutation rate) and if the occurrence of each mutation doubled the chance of another mutation occurring in the same cell, the probability that five simultaneous mutations would occur in any one individual would be 1 x 10^{22} (.0000000000000000000001). This means that if the population averaged 100,000,000 individuals and if the average generation lasted but one day, such an event as the appearance of five simultaneous mutations in one individual, would be expected once in every 274 billion years.[38]

Evidence for the evolution of *plants* is just as lacking as that for animals. C. A. Arnold stated:

> It must freely be admitted that this aspiration [of finding evidence for plant evolution] has been fulfilled to a very slight extent, even though

36. Ibid., p. 149. Cf. Francis Hitching, *The Neck of the Giraffe*, pp. 28–31.

37. Cf. R. E. D. Clark, *Darwin: Before and After* (Chicago: Moody Press, 1967), p. 126; Michael Pitman, *Adam and Evolution*, p. 64.

38. *Tempo and Mode in Evolution* (New York: Hafner Pub. Co., 1944), pp. 54 ff.; cited by John W. Klotz, *Genes, Genesis, and Evolution* (St. Louis: Concordia Pub. House, 1955), p. 298. Michael Pitman reminds us, "For evolution to occur through mutation, countless sequential good mutations would be required; at each step all would have to cooperate harmoniously and each mutation would have to be selected for. This simply could not happen" (*Adam and Evolution*, p. 67).

> paleobotanical research has been in progress for more than one hundred years."[39]

And what about *insects*? "We are in the dark concerning the origin of insects," says Pierre-P. Grassé, renowned French zoologist, who is a former President of the Acadamie des Sciences and editor of the thirty-five volume *Traite de Zoologie* (1948–72).[40]

A third very serious limitation to the potential for variation in the living world is the presence of highly complex organs and structures that cannot function effectively unless they are complete. "*They are either perfect or perfectly useless*." For example, *the human ear*,

> . . . is intricate beyond imagination. . . . The organ of Corti alone, a spiralling 3mm diameter ridge of cells in the inner ear that seems to play a crucial part in the way we hear pitch and direction of sound, contains some 20,000 rods and more than 30,000 nerve endings.[41]

How could the ear function at all if the separate parts had to come together by chance through millions of years?

And what about *the human eye*, with its 130,000,000 light-sensitive rods and cones? These ". . . cause photochemical reactions which transform the light into electrical impulses." Every second, *one billion of these impulses are transmitted to the brain*![42]

> Now it is quite evident that if the slightest thing goes wrong *en route*—if the cornea is fuzzy, or the pupil fails to dilate, or the lens becomes opaque, or the focussing goes wrong—then a recognizable image is not formed. The eye either functions as a whole, or not at all. So how did it come to evolve by slow, steady, infinitesimally small Darwinian improvements? Is it really possible that thousands upon thousands of lucky chance mutations happened coincidentally so that the lens and the retina, which cannot work without each other, evolved in synchrony? What survival value can there be in an eye that doesn't see? Small wonder that it troubled Darwin. 'To this day the eye makes me shudder,' he wrote to his botanist friend Asa Gray in February, 1860.[43]

39. *An Introduction to Paleobotany* (New York: McGraw-Hill Pub. Co., 1947), p. 7. Cited in Duane T. Gish, *Evolution: The Challenge of the Fossil Record* (El Cajon: Master Books, 1985), p. 232.

40. Pierre-P. Grassé, *Evolution of Living Organisms* (New York: Academic Press, 1977), p. 30.

41. F. Hitching, *The Neck of the Giraffe*, p. 90, 91.

42. Ibid., pp. 85–6.

43. Ibid., p. 86. See also Michael Pitman's critique of evolutionary explanations for the origin of the eye, in *Adam and Evolution*, pp. 215–18. A large section (pp. 261–323) of The

Horses—Products of Evolution?

It has frequently been claimed that fossil horses reveal a clear line of development from the small dog-sized specimens to the large types we see so frequently today. But there is no place in the world where a direct-line fossil sequence from small to large can be found! Therefore, different-sized creatures that had certain horse-like features could have lived contemporaneously in different parts of the world. It cannot be demonstrated that they were all members of the same kind; but even if they were (like the varieties of dogs), this does not prove that they evolved from small to large and from simple to complex. There are many subtle assumptions that underlie the interpretation of fossil finds, not the least of which are *uniformitarianism* (e.g., a gradual deposition of sediments and formation of fossils) and *evolutionism* (an increase of complexity through beneficial mutations). Both assumptions are contradicted by empirical science and an honest exegesis of Scripture. See Francis Hitching, *The Neck of the Giraffe* (New Haven and New York: Ticknor & Fields, 1982), pp. 28–31.

Where do ears and eyes come from then? The infinite, personal, living God of creation calls for our attention:

> The hearing ear and the seeing eye, the LORD has made both of them (Prov. 20:12).
>
> Pay heed, you senseless among the people; and when will you understand, stupid ones? He who planted the ear, does He not hear? He who formed the eye, does He not see? He who chastens the nations, will He not rebuke, even He who teaches man knowledge? The LORD knows the thoughts of man, that they are a mere breath (Ps. 94:8–11).

Controversy is now swirling around the obvious complexities of the *bombardier beetle*. *Time* (February 25, 1985, p. 70) noted that the bombardier beetle

> . . . does appear to be unique in the animal kingdom. Its defense system is extraordinarily intricate, a cross between tear gas and a tommy gun. When the beetle senses danger, it internally mixes enzymes contained in one body chamber with concentrated solutions of some rather harmless compounds, hydrogen peroxide and hydroquinones, confined to a second chamber. This generates a noxious spray of caustic benzoquinones, which explodes from its body at a boiling 212 F. What is more, the fluid is pumped through twin rear nozzles, which can be rotated, like a B-17's gun turret, to hit a hungry ant or frog with bull's eye accuracy.

However, any creationist implications of such astounding complexities are dismissed by the editor through an appeal to biologist Thomas Eisner of Cornell University. Eisner's conclusion: The bombardier beetle "just found novel uses for existing elements"![44]

The *duckbill platypus* has also gained notoriety as a battleground in the creation/evolution debate. "Found only in Australia, it lays eggs like a reptile yet has milk glands like a mammal, swims like a duck but has the fur of a beaver . . . Now, in a recent issue of the British journal *Nature*, scientists report that the platypus is plugged in: it can [with its bill]

National Geographic Society's magnificently illustrated volume, *The Incredible Machine* (1986) is devoted to an analysis of how the world reaches us through our five sense receptors (vision, hearing, smell, taste, touch). For the complexity of the human brain, see below, pp. 125–27.

44. For a careful refutation of Eisner's conclusion, see A. J. Monty White, "The Bombardier Beetle and Its Use in the Creation-Evolution Debate," *Creation News* 49 (3 Church Terrace, Cardiff CF2 5AW, United Kingdom, October, 1985), pp. 1–2.

sense electrical fields, the first higher vertebrate found to have that ability."[45]

But what is the author's conclusion concerning this fascinating creature?

> The electrical receptors of the platypus are so different from [sharks, skates, and rays] that the curious mammal probably evolved its electrical sense independently. If true, scientists have fresh support in their battle with creationists. The appearance of such an adaptation in unrelated species suggests that it is not a trait so startling that only divine intervention can explain it. Darwin's theory of mutation and natural selection would account for it. If so, the platypus' bill is more evidence that natural forces underlie the rich variety of the global zoo.[46]

Quite to the contrary, the "duckbill" of the platypus, together with its entire structure and functions, constitute a shattering blow to the credibility of evolutionism in all its forms.

Hundreds of additional examples could be provided from recent and careful studies of the biologic world.[47] One of the most familiar cases is the feathered wing of a bird. The majority of evolutionists today, still following the outmoded (see below) neo-darwinist views of Ernst Mayr, Theodosius Dobzhansky, Julian Huxley, and George Simpson, insist that birds evolved from reptiles through an accumulation of gradual adaptations. But how could this have happened? Are we to suppose that a few reptiles began to develop appendages on the sides of their bodies that grew in size and complexity through millions of years until they finally attained the power of flight? Even granting that a reptile could produce such structures, which is absurd from the standpoint of Mendel's laws, how could such creatures have survived in the struggle for existence? Natural selection would have eliminated them long before they could fly. Furthermore, the entire structure and instinct pattern of the animal would have to be changed to enable it to take off from the ground. The same problem applies to insects, pterodactyls, and bats. It was once thought that the Archaeopteryx was a link between reptiles and birds, because it had some features of both. But this creature was no more a

45. Sharon Begley, "A Plugged-in Platypus," *Newsweek* (February 17, 1986), p. 78.

46. Ibid.

47. See John W. Klotz' discussions of the obligate relationship between the Yucca moth and the Yucca plant, and between the commercial fig and the Blastophaga wasp, in *Studies in Creation* (St. Louis: Concordia Publishing House, 1985), pp. 205–07.

The Duck-Billed Platypus of Australia

So astounding is the combination of physical features in this animal that scientists in England, who first saw a dead specimen, believed that it had been sewn together by Chinese merchants to deceive the British! Their confusion was understandable. The platypus has a ducklike beak, five webbed toes, swims like a fish and lays eggs. Like a bird, it makes a grass-lined nest and hatches its eggs by curling up on the nest and warming them. It must therefore be some kind of a special bird.

On the other hand, the platypus has four legs, a fur hide, a large flat beaverlike tail and claws like many mammals. It must therefore be some kind of a special mammal!

However, when it is small it has teeth which in the adults are replaced by horny plates, unique among mammals. Furthermore,

> It uses echo location like a bat or dolphin; a hollow spur on the inside of its heel connects with a poison-gland, making it the only venomous furred creature. Its legs are as short as a reptile's but it has large cheek pouches like a monkey or squirrel (Michael Pitman, *Adam and Evolution,* 1984, p. 210).

Evolutionists simply cannot handle animals like the platypus, which have "weird" combinations of traits that do not fit the "family tree of life" concept of neo-darwinism or even saltationism. For the Designer and Creator of the world, however, the platypus is no problem at all! Each animal possesses characteristics through which God intended to display His wisdom to men.

As a former evolutionist (at Princeton University), I can now say to God in a new way each day, as I discover the wonders of His power,

> I know that Thou canst do all things . . . Therefore I have declared [in my years as an evolutionist] that which I did not understand, things too wonderful for me, which I did not know. . . . I have heard of Thee by the hearing of the ear; but now my eye sees Thee; therefore I retract, and I repent in dust and ashes (Job 42:3–6).

link between reptiles and birds than the duck-billed platypus is a link between mammals and birds. The Archaeopteryx had full wings and perfect feathers, and no evolutionist has succeeded in explaining where wings and feathers came from.[48]

It was because of problems like these, but especially because no links between major types have been found in the fossil record, that leading paleontologists such as Richard Goldschmidt, O. H. Schindewolf, and, more recently, Stephen Jay Gould of Harvard, finally abandoned neodarwinian gradualism and turned to *catastrophic macro-saltations* involving *massive bursts of cosmic radiation* that could somehow produce not only genetically "hopeful monsters," but matching monster-mates at the same time, here and there, totally by chance! Such a theory, called '*punctuated equilibrium*' by Gould, neatly solves the problem of missing links and natural selection; but from the standpoint of the science of genetics it is a true monstrosity, analagous, perhaps, to Fred Hoyle's Steady-state Theory in astronomy, whereby hydrogen atoms all over the universe were assumed to have simply sprung into existence out of nothing. Crushed by the obvious requirements of the first law of thermodynamics, Hoyle's theory has been almost universally abandoned.

Thus, in answer to the question of where the first birds came from, Schindewolf stated: "The first bird crept out of a (mutated) reptile's egg."[49] But if fully developed birds with wings and feathers could hatch directly out of the eggs of reptiles, what is to prevent one from imagining that the first marsupials and placental mammals and even human beings might have appeared in a similarly miraculous way? Creationists believe that only a personal and powerful God can perform miracles; but they also believe that the miracles He performed in creating the living world were *very* different from those envisioned by "saltationists" such as Goldschmidt and Gould.

In answer to this, it might be argued that the doctrine of creation is equally absurd as an alternative to micromutationism, for it involves the sudden appearance of full-grown birds out of inorganic matter. But this is not a valid objection, for creationism has a dynamic at its disposal that neither Goldschmidt nor Schindewolf nor Gould had any right to appeal

48. Cf. Michael Pitman, *Adam and Evolution*, pp. 218–27.

49. "Der erste Vogel kroch aus einem (abgewandelten) Reptilei," *Grundfragen der Palaontolgie*, (Stuttgart, 1950), p. 277; cited in R. Hooykaas, *The Principle of Uniformity* (Leiden: E. J. Brill, 1963), p. 128. See Sharon Begley's analysis in "Science Contra Darwin, *Newsweek* (April 8, 1985), pp. 80–81.

to—*the infinite power of the personal God of creation*! And without God, evolutionists have nothing else to fall back on to provide the necessary dynamic for their theories than the power of *magic*.

The Lord Jesus Christ and His apostles have taught us by both precept and example to accept on God's authority the absolute historicity (as well as the inerrancy) of pre-Abrahamic events as recorded in Genesis 1–11. *Every one* of these eleven chapters is referred to somewhere in the New Testament. Furthermore, *every New Testament writer* refers to Genesis 1–11. And finally, *the Lord Jesus Christ referred to each of the first seven chapters of Genesis in a way that presupposes its historical truth*. Thus, we may have perfect confidence that no discovery of "science" can contradict the clear teachings of God's written revelation.[50]

Evolutionism has not produced a credible model for explaining how billions of fossils of plants and animals were preserved in the crust of the earth. It did not require long ages for these strata to be deposited. Fossils in the geologic column do not speak of a sequence of isolated creation events (and certainly not of evolutionary processes!), but rather of the sequences of death and burial through the hydrodynamic complexities of the great Flood. Therefore, there is no objective geologic evidence to contradict the biblical record that fruit trees were created before marine creatures.[51]

Evolutionists have also experienced total failure in their efforts to demonstrate how the first life could have developed from lifeless chemicals.[52] Therefore, no empirical evidence is available to contradict the testimony of Genesis that God created vast numbers of plants and animals, both small and great, within just a few days. The single "tree of life"

50. Cf. J. C. Whitcomb, "Contemporary Apologetics and the Christian Faith" (the W. H. Griffith Thomas Lectures delivered at Dallas Theological Seminary in February, 1977), *Bibliotheca Sacra* 134:534 (four issues, beginning in April 1977). On the teachings of Christ and the apostles on biblical inerrancy, see Wayne A. Grudem, "Scripture's Self-Attestation," in D. A. Carson and John D. Woodbridge, eds., *Scripture and Truth* (Grand Rapids: Zondervan Publishing House, 1983), pp. 19–59.

51. Cf. Henry M. Morris, ed., *Scientific Creationism*, rev. ed. (El Cajon, CA: Master Books, 1985), pp. 101–30; Ian T. Taylor, *In The Minds of Men: Darwin and the New World Order* (TFE Publishing, P. O. Box 5015, Stn. F, Toronto M4Y 2T1, Canada, 1984), pp. 81–114, 282–339; and Henry M. Morris and Gary E. Parker, *What Is Creation Science?* (San Diego: Creation-Life Publishers, 1982), reviewed by John C. Whitcomb, *Grace Theological Journal* 4:1,2 (Spring and Fall, 1983), pp. 109–17, 289–96.

52. For documentation, see above, Note 21.

Birds

Charles Darwin and his "neo-Darwinist" disciples today have completely failed to explain the origin of birds. Such amazingly adapted creatures could *never* have come into existence by a gradual and chance accumulation of mutations in the bodies of certain reptiles. It is just as ridiculous to imagine that birds were once reptiles as it would be to imagine that airplanes could be produced by attaching wings to the sides of trucks. Flying birds are obviously *designed for flight,* and every minute aspect of their physical form and instinct pattern contributes to this marvellous capacity. Darwin pointed to the different varieties of finches on the Galapagos Islands as "Exhibit A" for evolutionism. But these variations occur within strict limits, and they are all still finches (see Walter E. Lammerts, "The Galapagos Island Finches," in *Why Not Creation?,* Presbyterian and Reformed Publishing Co., 1970, pp. 354–366). At present, there are almost 8,600 distinct species of birds in existence, but there were more in the beginning than there are now. (Note how the "kinds" of birds in Leviticus 11:13–19 closely match our taxonomic "species.") Living kinds can (and many already have) become extinct; but none can ever evolve. For additional evidence that flying creatures could not have evolved, see Michael Pitman, *Adam and Evolution* (1984), pp. 218–27; and Michael Denton, *Evolution: A Theory in Crisis* (1986), pp. 199–216.

concept of evolutionism has suffered staggering blows during the hundred years of post-darwinian search for transitional forms (connecting branches) that never existed. Thus, God's revelation in nature is increasingly being found to harmonize with God's revelation in Scripture: all plants and animals and men reproduce after their kinds, as a vast, complex, and spectacularly beautiful biologic forest of permanently separate "trees of life."

Scientists are obviously very far from achieving a total understanding of even the simplest phenomena of the universe. But if and when God permits men to discover some of these secrets they will be found to be in full harmony with the teachings of the Bible. Until that day, and while many questions and shadows still remain, God would have us take His Word for what He states it to be—man's only infallible guide to all truth.

Conclusion

God's written record of the creation of living, sub-human organisms on planet earth is in full harmony with His revelation in nature. Every kind of plant and animal speaks its message to sinful man: we are sustained in life moment by moment by the great God who designed and created us by His power and wisdom. Lilies of the field and sparrows in the air, as well as huge monsters of the sea and land, join together in a great chorus of praise to Him who made them suddenly and supernaturally according to a master plan that was "very good" and pleasing in His sight. Satan, our unseen enemy, would distort and hide this perspective from us through such blasphemies as organic evolutionism. But some day, thank God, ". . . the earth will be full of the knowledge of the LORD as the waters cover the sea" (Isa. 11:9), and false concepts concerning the origin and nature and ultimate purpose of plants and animals will vanish away forever.

> But now ask the beasts, and let them teach you; and the birds of the heavens, and let them tell you. Or speak to the earth, and let it teach you; and let the fish of the sea declare to you. Who among all these does not know that the hand of the LORD has done this, in whose hand is the life of every living thing, and the breath of all mankind? (Job 12:7–10).

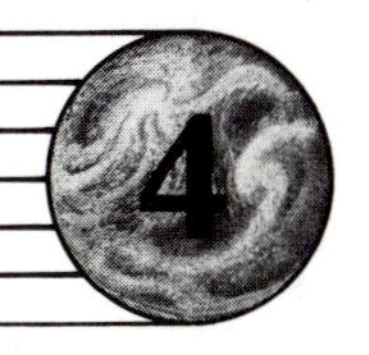

The Creation of Mankind

The Dignity of Man

As he tended his father's sheep by night and gazed into the heavens, David was overwhelmed by the magnitude of God's starry universe. Could a God of such power and transcendence have any real interest in these mere specks of cosmic dust called men? Astronomy as such could provide no comfort for David in this desperately crucial problem; and the enormous advances in astronomical knowledge we have experienced since his day still leave us in utter darkness. Modern astronomers, peering through gigantic telescopes, have yet to discover a single trace of the grace and love of God anywhere in the universe.

All true Christians would agree that the answer to *this* question must come from the inscripturated Word of God and from there *alone*. It was to the first chapter of Genesis that David appealed as his source of assurance that God created man a little lower than Elohim (the realm of deity) and crowned him with glory and honor, giving him dominion over all creation (Ps. 8:5–8; cf. Gen. 1:26–28). In spite of the conspicuous failure of *natural revelation* at this point, however, *special revelation* assures us that the human race *is* the object of God's loving concern, and one human being is more important to God than all the stupendous galaxies of the universe.

Looking carefully about him at a world groaning under the bondage of corruption, the brilliant author of the Book of Ecclesiastes saw no *empirical* basis for distinguishing human beings from beasts.

> I have seen under the sun that . . . the fate of the sons of men and the fate of beasts is the same. As one dies so dies the other; indeed, they all have the same breath and there is no advantage for man over the beast, for all is vanity. All go to the same place. All came from the dust and all return to dust (Eccles. 3:16–20).

Three thousand years later our scientific advances have not helped us at all in solving this problem. No one can prove experimentally that the spirit of a beast vanishes at death but that the spirit of man continues to exist forever. From the standpoint of chemistry, a good case could be made for the proposition that man is on the same level with animals. Both are made of the same "dust." Modern scientists, peering through powerful microscopes fail to see any trace of the image of God in the chemicals of man's body. All true Christians would agree that the final answer to *this* question also must come from the Bible and from there *alone*. Once again, the first chapter of Genesis is seen to be foundational to our faith, when natural revelation fails us.

But some Christians are not willing to carry this principle of the ultimate authority of Scripture to its logical conclusion. They acknowledge that the Bible, rather than astronomy and chemistry, is our source of information concerning the *dignity* of man. But they cannot bring themselves to believe that the Bible rather than physical anthropology is our source of truth concerning the *creation* of man. Here, at least, we are told, natural revelation has equal authority with special revelation in Scripture, and wherever there is conflict the early chapters of Genesis must be molded into the framework of contemporary scientific theory concerning the origin of man. One writer expressed it as follows:

> Both the Bible of nature and the Holy Bible are infallible, each in its own way, because both are written by the almighty hand of God; otherwise, speaking with all reverence, God could not be trusted. Evolution is not just about all hypothesis. We are compelled to believe that at least much of it is true. And we may not be silent about that. . . . It is the result of reading the Bible of nature directly.[1]

Another wrote:

> I view the opening chapters of Genesis as a poetic expression of the God-inspired author of that book. I believe that this part of Scripture should

1. Peter G. Berkhout, *The Banner* (March 5, 1965), p. 22.

> not be regarded as a scientific textbook. . . . Should not a Christian be permitted to use science to delve into the unsolved mysteries of creation?[2]

This viewpoint, which may be called the "double-revelation theory," maintains that God has given to man two revelations of truth, each of which is fully authoritative in its own realm: the revelation of God in Scripture and the revelation of God in nature. Although these two revelations differ greatly in their character and scope, they cannot contradict each other, since they are given by the same self-consistent God of truth. The theologian is the God-appointed interpreter of Scripture, and the scientist, we are told, is the God-appointed interpreter of nature. Furthermore, each has specialized tools for determining the true meaning of the particular book of revelation which he is called upon to study.

The Double-revelation Theory further maintains that whenever there is an apparent conflict between the conclusions of the scientist and the conclusions of the theologian, especially with regard to such problems as the origin of the universe, the solar system, the earth, plant and animal life, and man; the effects of the Edenic Curse; and the magnitude and effects of the Noahic Deluge; the theologian must rethink his interpretation of Scripture at these points in such a way as to bring the Bible into harmony with the general consensus of scientific opinion, since the Bible is not a textbook of science, and these problems overlap the territory in which science alone must give us the detailed and authoritative answers.

Advocates of the double-revelation theory hold that this is necessarily the case, for if an historical and grammatical interpretation of the biblical account of Creation, the Edenic Curse, the Flood, and the Tower of Babel should lead the Bible student to adopt conclusions that are contrary to the prevailing views of trained scientists concerning origins, then he would be guilty of making God a deceiver of mankind in these vitally important matters. But a God of truth cannot lie. Therefore Genesis must be interpreted in such a way as to agree with the generally-accepted views of modern science. After all, we are reminded, Genesis was written only to give us answers to the questions "Who?" and "Why?" but modern science must answer the important questions "When?" and "How?"[3]

2. Robert C. Homan, *The Banner* (October 22, 1965), p. 20.

3. Cf. John C. Whitcomb, *The Origin of the Solar System* (Phillipsburg, NJ: Presbyterian and Reformed Pub. Co., 1964), pp. 7–9, 25–30, for additional evaluation and documentation of the "double-revelation theory."

Theistic Evolution

Following this general approach to the early chapters of Genesis, a number of Christian men of science (and probably an even larger number of theologians following in their footsteps) have adopted the view that Adam's body was simply that of some animal which had providentially been evolved into a biped through millions of years of gradual changes until God put within it an eternal soul several hundred thousand years ago. Thus, in *Evolution and Christian Thought Today,* Walter Hearn and Richard Hendry concluded a chapter on "The Origin of Life" with these words: "The authors of this chapter consider the expressions of Scripture regarding the creation of life to be sufficiently figurative to imply little or no limitation on possible mechanisms."[4] In a symposium entitled "Origins and Christian Thought Today," held at Wheaton College on February 17, 1961, Dr. Hearn further clarified his position:

> . . . surely we know that processes have been involved in bringing *us* into existence. Why shudder, then, at the idea that processes were involved in bringing Adam into existence? Granted that we do not yet know details of the processes, why may we not assume that God *did* use processes?[5]

Henry W. Seaford, Jr., testified that

> An evolutionary view of man's place in nature assists my understanding of the antithesis between flesh and spirit. When I instruct my children in morality I can explain that the human body is an animal closely related to other Higher Primates. . . .[6]

Jan Lever of the Free University of Amsterdam agreed:

> When we thus place side by side the knowledge which we possess of the higher life of the Primates of the Pleistocene Epoch and the revelation that man has been brought forth within that which has been created, then we may not reject in advance the *possibility* that the genesis of man occurred by way of a being that, at least with respect to the characteristics of its skeleton, was an animal, according to our norms and criteria . . . we

4. *Evolution and Christian Thought Today,* ed. Russell L. Mixter (Grand Rapids: Eerdmans, 1959), p. 69.

5. *Journal of the American Scientific Affiliation* 14:2 (June, 1961), p. 42.

6. "Near-Man of South Africa," *Gordon Review* 4:4 (Winter 1958), pp. 187–189.

> may not reject in advance the possibility that there has existed a genetic relation between man and animal.[7]

Under pressure from these and similar statements by Christian men of science, Edward John Carnell, who served for several years as president of Fuller Theological Seminary, retreated to the following position:

> Since orthodoxy has given up the literal-day theory out of respect for geology, it would certainly forfeit no principle if it gave up the immediate-creation theory out of respect for paleontology. The two seem to be quite parallel. . . . If God was pleased to breathe His image into a creature that had previously come from the dust, so be it.[8]

Thus, natural revelation, as interpreted by "the modern scientist," cannot long remain on an equal footing with special revelation in Scripture, but finally overpowers and supersedes it.

Theistic evolution cannot consistently allow for any physical miracle in Adam's "creation." Thus, even after the image of God was put into a male and a female ape, their bodies, being unaffected by this spiritual miracle, would continue to be subject to disease and death just like the bodies of other apes. Therefore sin could not be the cause of physical death even in the human race, and Romans 5:12 would be incorrect when it tells us that "through one man sin entered into the world, and death through sin." . . .

Some theistic evolutionists have frankly recognized the theological implications of their surrender to evolutionary anthropology and have been willing to adjust their theology accordingly. Peter Berkhout, for example, stated:

> We realize full well, that, if what we call theistic evolution were accepted as true, a tremendous change would take place in our thinking; compared with which the change to the Copernican point of view would be a mere bagatelle. For example, if man did descend from some primate physically, can we attribute all imperfection and all of what we call physical evil to man's Fall? Is it not an oversimplification anyway? Many of our books

7. *Creation and Evolution* (Grand Rapids: Grand Rapids International Publications, 1958), pp. 197, 221.

8. Edward John Carnell, *The Case for Orthodox Theology* (Philadelphia: Westminster Press, 1959), p. 95. See also Harold J. Ockenga, *Women Who Made Bible History* (Grand Rapids: Zondervan, 1962), p. 12.

Mankind

Man is the crown of God's creation. He was made in the image and likeness of his Creator and was given complete dominion over the earth (Gen. 1:26). "The heavens are the heavens of the LORD; but the earth He has given to the sons of men" (Ps. 115:16). Fallen man has lost that original dominion, but still possesses God's image (Gen. 9:6; James 3:9). Redeemed through Christ, God's incarnate son, believing men have already been moved positionally from the realm of "little lower than angels" (Heb. 2:7) to a realm "far above all rule and authority and power and dominion, and every name that is named . . ." (Eph. 1:21; cf. 2:6). Glorified men will even judge angels (1 Cor. 6:3).

In the light of all this, how utterly blasphemous is the currently popular idea that man is little more than "a naked ape." *The physical differences* between men and apes are enormous, as careful observation clearly shows. But if the physical differences are great, *the mental-cultural-spiritual differences* are little short of infinite. Of all living beings on this planet, only man is self-conscious as a person; is sufficiently free from the bondage of instinct to exercise real choices and to have significant purposes and goals in life; has complex emotions including sadness and joy; appreciates art and music creatively; can make real tools; can be truly educated rather than merely trained; can use oral and written symbols to communicate abstract concepts to other persons and thus enjoy true fellowship; can accumulate knowledge and attain wisdom beyond previous generations and thus make genuine history; can discern moral right and wrong and suffer agonies of conscience; can recognize the existence and rightful demands of his Creator through worship, sacrifice, and religious service. Only man will exist forever as a personal being either in heaven or in hell.

would have to be rewritten. But if necessary for the sake of truth, why not?[9]

Charles Darwin also traveled down this path. "First of all his faith in the Old Testament was shattered. Then he could no longer believe in the miracles of the New. Finally he was left wondering whether Christianity was a Divine revelation at all."[10]

In order to avoid such total theological disaster, Pope Pius XII, in his encyclical *Humani Generis* (1950), took refuge in a more modified form of theistic evolution. While warning Roman Catholic scientists and theologians to exercise "the greatest moderation and caution in this question," and to weigh and judge various opinions "with the necessary seriousness, moderation and measure," in view of the Church's commitment to the unity of mankind in Adam and the historicity of his original sin, he nevertheless gave them "liberty" to research and discuss "the doctrine of evolution, in as far as it inquires into the origin of the human body as coming from pre-existent and living matter."[11]

The inconsistency of this position is obvious. In order for God to change a mortal ape into a sinless and therefore immortal man, who had sufficient vigor to live 930 years beyond the Fall and the Curse, a physical as well as a spiritual miracle had to occur. But if one allows such a *physical* miracle in the creation of Adam in order to preserve some of the essentials of Christianity, by what form of logic does one then deny the *physical* miracle of the direct creation of Adam's body, which is clearly taught in the New Testament as well as in the second chapter of Genesis?[12]

In spite of the warnings included in the papal encyclical, many Roman Catholic theologians have clearly allowed for the evolution of man's body from animals: "If ever it should be established [that the human body has

9. "Revelation and Evolution," from a Netherlands publication cited by John Vander Ploeg, *The Banner*, October 8, 1965, p. 9.

10. R. E. D. Clark, *Darwin: Before and After*, p. 83.

11. Articles 36, 37. Cf. Claudia Carlen, *The Papal Encyclicals* 1939–58 (Raleigh: McGrath Publishing Co., 1981), pp. 181–82. J. F. Ewing, in the Roman Catholic *Anthropological Quarterly* 29 (October, 1956), stated: "There is no officially proclaimed doctrine of the Catholic Church which is in contradiction with a theory of the evolution of Man's body" (p. 123).

12. Davis A. Young, a Christian geologist who is committed to the basic validity of the evolutionary timetable of earth history, and even to the evolution of the plant and animal kingdoms, nevertheless states, "The idea of theistic evolution of man is an unbiblical idea and ought to be abandoned by those who profess to believe in a genuinely Biblical Christianity" (*Creation and the Flood*, p. 144).

been evolved from lower forms], the religious teaching of Genesis would remain the same . . ."[13]

The Direct Creation of Adam's Body

For those who are willing to search the Scriptures and to believe what they say, nothing can be clearer than the fact that God directly created the bodies of Adam and Eve wholly apart from the use of previously existing animals.

Let us begin with the New Testament. When the Pharisees confronted the Lord Jesus Christ with the divorce question (Matt. 19:3), He answered them by confirming the permanence of the marriage bond in terms of Genesis 2:24. It is important to observe that our Lord introduced this reference to the first marriage by appealing to the *physical* basis for it: "Have you not read, that He who created them from the beginning MADE THEM MALE AND FEMALE . . . ?" (Matt. 19:4; cf. Gen. 1:27). Thus the Lord Jesus Christ clearly confirmed the teaching of Genesis that God created Adam and Eve, not only in His image and likeness (spiritually), but also as male and female (physically). If Adam and Eve were animals before they received God's image and likeness, they would *already* have been male and female, thus rendering the statements of Genesis 1:27 and Matthew 19:4 both inaccurate and misleading.

The apostle Paul clearly agreed with this concept of the physical uniqueness and thus the supernatural origin of the human race when he wrote: "All flesh is not the same flesh, but there is one flesh of men, and another flesh of beasts . . ." (1 Cor. 15:39). One of the basic thrusts of theistic evolution, of course, is that all flesh *is indeed* "the same flesh," the human race being a mere twig on the branch of anthropoid mammals. Paul's statement is in obvious contradiction to this theory.

Another very clear New Testament statement concerning the supernatural origin of mankind is found in 1 Corinthians 11:8, 12—". . . man does not originate from woman, but woman from man. . . . For as the woman originates from the man, so also the man has his birth through the woman. . . ." Paul is quite obviously stating that while all men (and women) today have mothers, all women (and men) had their ultimate origin in a man (Adam); because, as he had announced earlier to the "men of Athens" at Areopagus, God "made from one [i.e., Adam], every

13. E. F. Sutcliffe, "Genesis," in *A Catholic Commentary on Holy Scripture* (New York: Thomas Nelson and Sons, 1953), p. 185. Cf. Ian T. Taylor, *In the Minds of Men*, pp. 372–77.

nation of mankind . . ." (Acts 17:26). God accomplished this, of course, by Adam through Eve, who was therefore "the mother of all the living" (Gen. 3:20). But all of these statements can only be true if theistic evolution is false, for otherwise the first woman would have come physically from a female animal, not from a human male.

Turning now to the Old Testament, we come to the crucial text on man's physical creation, Genesis 2:7—"Then the LORD God formed man of dust from the ground, and breathed into his nostrils the breath of life; and man became a living being." To the extent that theistic evolutionists seriously interact with the biblical text at all, they insist that the "dust from the ground" from which God formed Adam was *living* "dust," namely, man's animal ancestry! Furthermore, we are told that Adam's creation simply involved the impartation of a *spiritual* nature to a subhuman creature, for the King James Version of Genesis says that "man became a living *soul*."

But consistent biblical hermeneutics does not permit Genesis 2:7 to be handled in this way! One of the basic laws of this time-honored and God-honored science is the *law of context*. According to this law, each passage of the Bible must be understood in the light of the passages that precede and follow it and ultimately in the light of the entire Bible. Otherwise, a passage could be twisted out of context and be made to teach something that it was never intended to teach. Basically, this is how every heresy and cult throughout church history sprang forth.[14]

Now the context of Genesis 2:7 demonstrates, in the first place, that the Hebrew phrase translated "man became a living soul" in the King James Version does not allow for a prehuman form of life for Adam's body. The phrase "living soul" (nêpêš ḥayâh) should actually be translated "living creature," or "living being" (NASB, NIV), for the same phrase appears in Genesis 1:20, 21, and is applied to sea creatures! In other words, the purpose of Genesis 2:7b is not to tell us that Adam had a unique soul (which we already learn by implication in Genesis 1:26, 27), but that Adam was not *any* kind of a living creature until he *became* one by the creative breath of God. Until that moment, he was inanimate, lifeless matter. The significance of this fact can hardly be overestimated.

This leads us to a second important discovery from a study of the context, namely, that "dust of the ground" cannot be understood sym-

14. Cf. James M. Sire, *Scripture Twisting* (Downers Grove, IL: InterVarsity Press, 1980), pp. 52–58.

bolically of animals but must be interpreted literally. Note, for example, the terms of God's curse upon Adam in the following chapter:

> Cursed is the ground because of you. . . . Both thorns and thistles it shall grow for you . . . by the sweat of your face you shall eat bread, till you return to the ground, because from it you were taken; for you are dust, and to dust you shall return (Gen. 3:17b–19).

Two interesting things are said here about the "ground" and the "dust" from which Adam was taken: (1) it would bring forth thorns and thistles, and (2) Adam would return to it. Now if "dust from the ground" symbolizes the animal kingdom in Genesis 2:7, what does it mean here? Does this passage mean that animals brought forth thorns and thistles as a result of the curse? And does it mean that Adam had to return to the animal kingdom when he died? Those who believe in reincarnation might favor the idea that "dust" here includes the animal kingdom, but a theistic evolutionist would hardly want to use this as a prooftext for his "living dust" concept! Thus, the hermeneutical law of context demands that "dust from the ground" in Genesis 2:7 be interpreted literally, and it completely excludes the possibility of an animal ancestry for man.

The second chapter of Genesis also makes it perfectly clear that Eve was taken physically, literally, and supernaturally from the side of Adam. If this point be granted, then the whole purpose of trying to interpret Adam's creation in evolutionary terms falls apart. To connect Adam's body with the animal kingdom but to admit that Eve's body was directly created would be absurd, either from the standpoint of evolutionary science or biblical creationism. We may not know in exact detail how God fashioned the bodies of our first parents, but that He created them miraculously and suddenly is the plain teaching of Scripture.

The Marvelous Complexity of the Human Body

The very design and structure of the human body demands a special and distinct origin, genetically unrelated to the animal kingdom. Man's *brain*, for example, is an incomparable marvel of complexity in the entire physical universe. Whereas electronic computers can store and recall billions of bits of information, the capacity of the human brain seems to be almost *infinite*. Furthermore, it continues functioning night and day for many years, carrying on a vast number of subconscious functions, sorting out auditory, visual, olfactory, gustatory, and tactile information, processing it, and enabling its owner to act upon it. Also, it can think about itself![15]

Scientists who apparently have no creationist commitments or presuppositions whatsoever are increasingly using biblical terminology to describe the human brain and body: "miracle," "wonder," "marvel," "creation," "design," etc. In a promotional letter for the September 1979 issue of *Scientific American*, W. H. Yokel waxed poetic and almost ecstatic on this subject:

> The deep new knowledge about the brain, gathered at an accelerated rate in recent years, shows this organ to be *marvelously designed and capacitated* beyond the wonders with which it was invested by ignorant imagination.
>
> Microelectronics can pack about a million circuits in a cubic foot, whereas the brain has been estimated to pack a million million circuits per cubic foot. Computer switches interact with not more than two other switches at a time, whereas a brain cell may be 'wired' to 1,000 other cells on both its input and output sides. . . .
>
> Perhaps the most elusive questions surround the brain functions that most make us human—the capacities of *memory* and *learning*. Transcending what might be called the hardware of the brain, there comes a software capacity that *eludes hypothesis*. The number that expresses this capacity in digital information bits *exceeds the largest number to which any physical meaning can be attached* [italics added].

In a promotional folder (February 1986), describing their new book about the human body, *The Incredible Machine*, The National Geographic Society invited potential readers "to journey inside the human body . . . the most miraculous creation of all." With 375 pages of spectacular photographs and paintings, the argument is overpowering: man is marvelously designed! But the foreword by Lewis Thomas and the opening chapter ("The Cosmic Creature") by Susan Schiefelbein assure us that man is nothing but "incredible magic" (p. 7) and "a celestial accident" (p. 11).

Can a mere accidental combination of mass, energy, chance and time produce a "miraculous creation" or "marvelously designed and capacitated" organs? By inspiration, David wrote words three thousand years ago which cannot be improved upon:

15. *Time*, January 14, 1974. "The mind can store an estimated 100 trillion bits of information—compared with which a computer's mere billions are virtually amnesiac" (*Newsweek*, September 29, 1986, p. 48).

> For Thou didst form my inward parts; thou didst weave me in my mother's womb. I will give thanks to Thee, for I am fearfully and wonderfully made; wonderful are Thy works, and my soul knows it very well (Ps. 139:13–14).[16]

While marveling at the enormously complex "super-computer" which resides inside the human skull, one must also agree with Wilder Penfield (*The Mystery of the Mind* [Princeton: Princeton University Press, 1975], p. 47):

> It is the mind that must first program the computer brain, since the computer is only a thing and, on its own, has no ability to make totally new decisions for which it is not programmed. . . . Man *has* a computer, not *is* a computer . . . and to treat him as a computer is like saying that a love letter should be the sole object of one's affection—not the sender.[17]

Ape-Men and Cave Men

The physical differences between men and apes are truly enormous, and can only be explained by the fact that their DNA coding, the original "blueprinting," was different.

> The human nose has a prominent bridge and an elongated tip which is lacking in the apes. . . . Man has red lips formed by an outrolling of the mucous membrane which lines the inside of his mouth; apes do not have this. Apes have thumbs on their feet as well as on their hands. . . . Man has the greatest weight at birth in relation to his weight as an adult. Yet at birth he shows the least degree of maturation and is by far the most helpless of creatures. Man's head is balanced on top of his spinal column; the head of the ape is hinged at the front instead of on top."[18]

16. For some fascinating perspectives on the marvels of the human body, see Paul Brand and Philip Yancey, *Fearfully and Wonderfully Made* (Grand Rapids: Zondervan Publishing House, 1980); reviewed by John C. and David C. Whitcomb in *Grace Theological Journal* 2:2 (Fall, 1981), pp. 333–39.

17. Arthur C. Custance, *The Mysterious Matter of the Mind* (Grand Rapids: Zondervan Publishing House, 1980), pp. 65, 66, 92.

18. Paul A. Zimmerman, *Darwin, Evolution and Creation* (St. Louis: Concordia Publishing House, 1959), p. 128. Michael Pitman expands on the differences in brain volume, hands and feet, pelvis and gait, teeth, face and jaw, language and literacy, in *Adam and Evolution* (1984), pp. 241–55.

The National Geographic Society has sponsored several expeditions to East Africa in the search for transitional forms between apes and men. Louis and Mary Leakey, their son Richard, and Donald Johanson have been among the most prominent in discovering fossils of creatures they have claimed to be ancestral to man, living over three million years ago. The cover of the November 1985 issue of their popular and influential magazine consisted of a holographic image of a supposed ape-man skull.

But various leading evolutionists are far from convinced that these "australopithecine" (i.e., "southern ape") fossils represent early stages of human evolution. Challenging early claims that these four-foot-high creatures walked upright like human beings, Solly Zuckerman, noted British anatomist, and Charles Oxnard, professor of anatomy at the University of Southern California Medical School, have concluded that they ". . . did not walk erect but they had a mode of locomotion that may have been similar to that of orangutans."[19]

Famous "ape-man" evidences, such as Piltdown, Nebraska, Neanderthal, Pithecanthropus Erectus, Peking, Homo Habilis, and Zinjanthropus Man, have already been discredited by careful and objective analysts. They are the remains either of animals or of true human beings, but not a transition between them. Furthermore, skulls that are truly human have been found buried in strata far below those that contain the supposed remains of their evolutionary ancestors. Thus, evolutionism emerges once again as a faith-commitment rather than objective, empirical science.[20]

Over a hundred years of careful search for transitional forms ("links") between ape-like animals and men have produced a very few (few enough to fit into a small cemetery!) and very questionable specimens. But if men have been evolving from animals for *millions* of years, where are all the human beings who should be here by now, to say nothing of their remains?

> The average family size today, worldwide, is about 3.6 children, and the annual population growth rate is 2 percent. . . . It is essentially incredible

19. Duane T. Gish, *Evolution: The Challenge of the Fossil Record*, p. 156; cf. pp. 149, 56, 59, 62.

20. Cf. Duane T. Gish, *Evolution: The Challenge of the Fossil Record*, pp. 130–228; Henry M. Morris, ed., *Scientific Creationism*, 2nd ed., pp. 171–202; M. Bowden, *Ape-Men: Fact or Fallacy?*, 2nd ed. (Bromley, Kent: Sovereign Publications, 1981); and John N. Moore, *How To Teach Origins* (Milford, MI: Mott Media, Inc., Publishers, 1983), pp. 185–265.

> that there could have been 25,000 generations of men with a resulting population of only 3.5 billion. If the population increased at only 1/2 percent per year for a million years, or if the average family size were only 2.5 children per family for 25,000 generations, the number of people in the present generation would exceed 10^{2100} a number which is, of course, utterly impossible: only 10^{130} electrons could be crammed into the entire known universe.[21]

In addition to all of these considerations are the heavy doubts concerning absolute dates assigned to "ape-man" fossils by investigators who are already committed to a geologic timetable of over two billion years. Kenneth F. Weaver, in a *National Geographic Magazine* feature article, "The Search For Our Ancestors" (November 1985), assures his readers that the vast ages being assigned to fossilized bones are based securely on the rate of decay of radioactive potassium to argon:

> It was the early use of the potassium-argon technique in 1961 to date the lowest level at Olduvai Gorge in Tanzania that radically lengthened the known time span of hominid evolution and ignited the explosion of knowledge about early man (p. 589).

But the unprovable assumptions that underlie this and other such techniques have been repeatedly and effectively exposed by highly qualified scientists for many years.[22]

Genetically and thermodynamically, of course, it is utterly impossible for an animal to become a human being. The fixed parameters of the enormously complex DNA code, the damaging effect of random mutations on the information stored in that code, and thus the downward "drift" or "drag" of genetic quality in living systems from generation to generation as predicted by the second law of thermodynamics all serve to destroy the credibility of organic evolutionism as a scientific hypothesis.

A Dutch zoologist, J. J. Duyvené de Wit, clearly demonstrated that the process of speciation (such as the appearance of many varieties of dogs and cats) is inevitably bound up with genetic depletion as a result of natural selection. When this scientifically established fact is applied to

21. Henry M. Morris, ed., *Scientific Creationism*, 2nd ed., pp. 167–69; see also Ian T. Taylor, *In The Minds of Men*, pp. 440–41.

22. Cf. Duane T. Gish, *Evolution: The Challenge of the Fossil Record*, pp. 51, 91; Henry M. Morris, *Scientific Creationism*, pp. 145–48.

the question of whether man could have evolved from ape-like animals, ". . . the transformist concept of progressive evolution is pierced in its very vitals."[23]

The reason for this, J. J. Duyvené de Wit went on to explain, is that the whole process of evolution from animal to man

> . . . would have run against the gradient of genetic depletion. That is to say, . . . *man [should possess] a smaller gene-potential than his animal ancestors*! Here, the impressive absurdity becomes clear in which the transformist doctrine [i.e., the theory of evolution] entangles itself when, in flat contradiction to the factual scientific evidence, it dogmatically asserts that man has evolved from the animal kingdom![24]

In the light of considerations such as these, which have been known for many years now, what must be said of Christians who surrender the direct creation of Adam and Eve out of respect for the currently prevailing views of evolutionary scientists?

No animal, whether ape, gorilla, chimpanzee, parrot, or dolphin, has ever uttered a rational word. Sir John Eccles observed,

> Experiments with chimpanzees who "talk" in sign language show that they can signal for things and get them, but "they don't *describe*. . . . They don't argue. . . . They have no value system. They don't make moral decisions. . . . They don't know they're going to die. . . . We must never judge animals as if they were just badly brought up human beings."[25]

The reason for this is biblically very clear: only man possesses the image and likeness of God, which includes rationality (Gen. 1:26; 5:1; 9:6; 1 Cor. 11:7; James 3:9). Thus, while the physical differences between men and primates are quite great, *the spiritual/mental/linguistic/cultural differences are little short of infinite.*

23. A *New Critique of the Transformist Principle in Evolutionary Biology* (Kampen, Netherlands: Kok, 1965), p. 56.

24. Ibid., p. 57.

25. "Photons, Philosophy, and Eccles," *The Washington Post* (March 15, 1981), p. F–1, quoted in Carl F. H. Henry, *God, Revelation and Authority*, 6:203. See also the brief but insightful analysis in *Time* (March 10, 1980), pp. 50–51; and Clifford Wilson and Donald McKeon, *The Language Gap* (Grand Rapids: Zondervan Publishing House, 1984). Even the National Geographic Society has modified some of its absurd claims in an article by Francine Patterson, "Conversations With a Gorilla" 154:4 (October, 1978), pp. 438–65 (cf. their volume, *The Incredible Machine* [1986], p. 356).

Of all living beings on this planet, *only man* is self-conscious as a person. *Only man* is sufficiently free from the bondage of instinct to exercise real choices and to have significant purposes and goals in life. *Only man* possesses an emotional capacity for sadness and joy. *Only man* appreciates art and music creatively. *Only man* can imagine and manufacture real tools. *Only man* can be truly educated rather than merely trained. *Only man* can use oral and written symbols to communicate abstract concepts to other persons and thus enjoy true fellowship. *Only man* can accumulate knowledge and attain wisdom beyond previous generations and thus make genuine history and experience progress. *Only man* can discern moral right and wrong and suffer the agonies of an offended conscience. *Only man* can be rightfully held accountable for his deeds, reckoned guilty, and judged. *Only man* can recognize the divine authority of his Creator and honor Him through worship, praise, prayer, sacrifice, and obedient service.

The concept of "ape-man" is therefore just as absurd, biblically and scientifically, as "giraffe-man" or "rhinoceros-man." But it is not merely an absurdity or an idle and harmless speculation of science-fiction. It is also a deadly perversion, for it strikes at the very image and likeness of God in man.[26]

There never has been an "ape-man"; but there are and have been and yet will be many millions of "cave men." Four thousand years ago Job described people in the regions of Northern Arabia,

> . . . who gnaw the dry ground by night in waste and desolation, who pluck mallow by the bushes, and whose food is the root of the broom shrub. . . . they dwell in dreadful valleys, in holes [caves] of the earth and of the rocks (Job 30:3–6).

And why did they live under such conditions? Because, "They were banished from their fellow men, shouted at as if they were thieves" (v. 5, NIV).

Tragic indeed has been the fate of such victims of human cruelty throughout human history. Possessors of the image of God, they nevertheless, in this fallen world, have been forced to live like animals, as hunters and food gatherers, never far from starvation. David, though a man after God's own heart (1 Kings 11:4; 15:3) was forced, with hundreds of his followers, to live in the caves of southern Palestine while King Saul

26. Cf. Carl F. H. Henry, *God, Revelation and Authority,* Vol. 6, p. 177.

and his armies sought to destroy him (1 Sam. 22–24). It was while he was a "cave man" that he composed two of his great psalms (Pss. 57, 142).

Indeed, many of God's people ". . . went about in sheepskins, in goatskins, being destitute, afflicted, ill-treated (men of whom the world was not worthy), wandering in deserts and mountains and caves and holes in the ground" (Heb. 11:37–38). Utterly incredible though it may seem, the Son of God Himself, through whom the universe was created (John 1:1–3), was rejected by His own nation (John 1:10, 11) and said: "The foxes have holes . . . but the Son of Man has nowhere to lay His head" (Matt. 8:20).

At the end of this age, the vast majority of human beings will be reduced to cave-man existence. "And the kings of the earth and the great men and the commanders and the rich and the strong and every slave and free man [will hide] themselves in the caves and among the rocks of the mountains" because of God's great judgments upon a world that has rejected His beloved Son and His gracious salvation (Rev. 6:15–17; cf. Isa. 2:19–22).

The Antiquity of Man

If natural revelation is superseded (but not contradicted) by special revelation with regard to man's *true dignity* and his *supernatural creation*, does it not at least provide for us the basis for determining his *antiquity*? James Oliver Buswell III, a Christian anthropologist who acknowledged the direct creation of Adam as the clear teaching of Scripture, believed,

> The creationist may accept the evidence for the age of pre-historic man and his culture. He need have no quarrel with an antiquity of hundreds of thousands of years; there is nothing in the Bible to indicate how long ago man was created.[27]

When Carl F. H. Henry criticized Christian anthropologists for giving way to "the inordinate pressures of contemporary scientific theory about the antiquity of man,"[28] Buswell replied that he and other anthropologists were simply following William Henry Green and Benjamin B. Warfield in their denial that the genealogies of Genesis placed *any* limitations on the antiquity of man.

27. *Evolution and Christian Thought Today*, p. 181.

28. Editorial, *Christianity Today* (January 15, 1965), p. 28. For his more recent analysis, see Carl F. H. Henry, "The Origin and Nature of Man," in *God, Revelation and Authority* (1983), 6:197–228.

> I'm sure that we Christian anthropologists would be willing to study with an open mind any serious, scholarly attempt to invalidate, overthrow, or supersede the classic works in this area upon which our position partially rests.[29]

In response to this challenge I would like to suggest several biblical limitations upon the antiquity of the human race. In the first place, to stretch the genealogies of Genesis 5 and 11 to cover a period of over a hundred thousand years is to do violence to the chronological framework of all subsequent Bible history. By means of biblical analogies, it is indeed possible to find gaps, especially in the genealogy of Genesis 11. But those very analogies serve to limit our time scale for Genesis 11. The gap between Amram and Moses was three hundred years, not thirty thousand (cf. Exod. 6:20; Num. 3:17–19, 27–28). And the gap between Joram and Uzziah in Matthew 1:8 was fifty years, not five thousand.[30]

In the second place, only three of the ten patriarchs listed in Genesis 11—Reu, Serug, and Nahor—are available for spanning the vast period of time demanded by these anthropologists, for the patriarchs listed before them preceded the Tower of Babel judgment and the scattering of mankind (cf. Gen. 10:25). And yet the clearest suggestion of a time gap in Genesis 11 occurs *before* this judgment, between Eber and Peleg, because of the sudden drop of average life span.[31]

In the third place, it is impossible to imagine that Reu, Serug, and Nahor, to say nothing of Lamech, Noah, and Shem, were savage, illiterate cave-dwellers of the stone-age period. The fourth chapter of Genesis, with its clear indication of cultural achievement, including the forging of "all implements of bronze and iron" (vs. 22), and Genesis 6, with its account of the great ark-building project, make such a theory completely untenable. Or are we to suppose that in some tiny pocket of civilization, nearly swamped by an ocean of human savagery, an unbroken chain of saintly men (some of whom lived for centuries) perpetuated the Messianic line of Shem and handed down the knowledge of the one true God for hundreds of thousands of years? Even to ask such a question is to answer it.

Finally, we must ask how certain details of the story of the great Flood could have been handed down from one primitive stone-age culture to another, purely by oral tradition, for hundreds of thousands of years, to

29. Letter to the Editor, *Christianity Today* (March 12, 1965), p. 22.
30. Cf. J. C. Whitcomb and H. M. Morris, *The Genesis Flood*, pp. 485–86.
31. Ibid., pp. 481–83.

be finally incorporated into the Gilgamesh Epic of the Babylonians? That such could have happened for several thousand years is remotely conceivable. That it could have happened over a hundred thousand years is quite inconceivable. The Gilgamesh Epic alone, rightly considered, administers a fatal blow to the concept of a vast antiquity for Adam and Noah.[32]

It is true that Benjamin B. Warfield (1851–1921) was one of the greatest orthodox theologians of modern times. But it is also true that he was capable of making mistakes. One such mistake, I believe, was his assertion that "two thousand generations and something like two hundred thousand years may have intervened" between Adam and Noah as far as the Scriptural data in Genesis 5 and 11 are concerned.[33] It should also be pointed out, however, that Warfield went on to say (and this statement is seldom quoted today) that man has probably not existed on earth more than ten to twenty thousand years.[34] It is unfortunate that one of Warfield's statements with regard to Genesis 5 and 11 is so frequently appealed to as the final word on the subject of pre-Abrahamic chronology.

It is of the utmost importance to recognize that while the absolute inerrancy of the biblical autographs (which B. B. Warfield clearly affirmed) is indeed *essential*, it is not *sufficient* as a basis for true theology in general or for true creationism in particular.[35] How one interprets the inerrant text of Scripture (i.e., the science of biblical hermeneutics) is the ultimate issue today, just as it was in Jesus' day. The Jewish leaders who opposed our Lord were in formal agreement with Him concerning the final authority of the Old Testament, but rejected His interpretations of the text.

32. Ibid., pp. 483–89.

33. "Antiquity and Unity of the Human Race" in *Biblical and Theological Studies* (Phillipsburg, N.J.: Presbyterian and Reformed Pub. Co., 1952), p. 247.

34. Ibid., p. 248.

35. In Article XII of the Chicago Statement on Biblical Inerrancy, for example, we find an *essential* but *insufficient* foundation for all studies of Genesis 1–11: "We deny that biblical infallibility and inerrancy are limited to spiritual, religious, or redemptive themes, exclusive of assertions in the fields of history and science. We further deny that scientific hypotheses about earth history may properly be used to overturn the teaching of Scripture on creation and the flood." This is a worthy statement as far as it goes; but many day-age concordists could endorse it without hesitation.

Thus, for concordists such as Davis A. Young[36] and theistic evolutionists such as David N. Livingstone[37] to build their case upon the theological compromises and concessions of various theologians of the post-darwinian era (such as Warfield) is to build upon a foundation of sand. These theologians would probably not have compromised their theological convictions on ultimate origins were it not for the immense pressure placed upon them by the uniformitarian/evolutionary consensus of the scientific establishment of the late nineteenth century. They naively assumed that scientists spoke of objectively verifiable realities when they spoke of the vast antiquity of the earth and the evolution of living things.[38]

Here we find a startling case of "Galileo in reverse." That great scientist of the early seventeenth century was not rejecting biblical revelation when he rejected the idea that the sun revolves around the earth. Rather, he was rejecting the views of the Latin "church" that had naively and prematurely adopted the perspectives of certain non-Christian (Aristotelian and Ptolemaic) "scientists."[39] So also today, Christians are in constant danger of prematurely adopting currently popular secular philosophies, only to discover, with great embarrassment, that they have married (and have forced the Scriptures to marry) the wrong system. Inevitably, like Jacob, they will awaken "in the morning" and find that, "behold, it was Leah!" (Gen. 29:25: cf. Rom. 12:1–2; Col. 2:8).

One of the disturbing trends in the Church today is the widespread refusal of Christian men of science to challenge the uniformitarian assumptions which underlie various schemes for determining the age of fossils. Why must Christians accept the timetable of evolutionary paleontology and anthropology when it involves a denial of the biblical doctrines of a supernatural creation-week, the entrance of physical death

36. Davis A. Young, *Christianity and the Age of the Earth*, pp. 58–59.

37. David N. Livingstone, "B. B. Warfield, the Theory of Evolution and Early Fundamentalism," *The Evangelical Quarterly* 58:1 (January, 1986), pp. 69–83.

38. For additional discussion of the response of early Princeton theologians to Darwinism, see my review of *The Post-Darwinian Controversies* by James R. Moore (Cambridge, England: Cambridge University Press, 1979) in *Grace Theological Journal* 2:1 (Spring, 1981), pp. 135–36.

39. For an excellent discussion of 'The Galileo Affair,' see Ian T. Taylor, *In the Minds of Men*, pp. 22–25. Cf. John C. Whitcomb, *The World That Perished* (Grand Rapids: Baker Book House, 1973), pp. 137–38.

The Fall to the Flood

(Dates are years after creation)

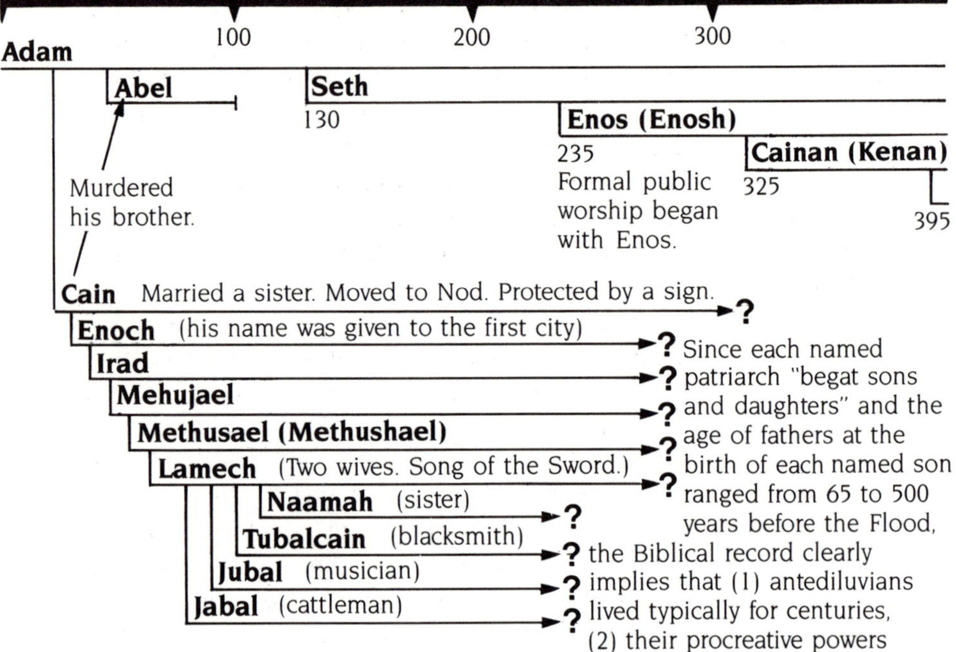

Since each named patriarch "begat sons and daughters" and the age of fathers at the birth of each named son ranged from 65 to 500 years before the Flood, the Biblical record clearly implies that (1) antediluvians lived typically for centuries, (2) their procreative powers persisted also for centuries, and (3) through the combined effects of long lives and large families, the earth could have been "filled" with people by the time of the Flood (Gen. 1:28; 6:1, 11, 13). This in turn implies that not many thousands of years elapsed between Creation and the Flood or between the Flood and the present.

---

into the world at the time of Adam's fall, and also the catastrophic effects of a geographically universal Flood in the days of Noah? These doctrines of Scripture are supported by many and varied arguments; and yet we find many evangelical scientists ignoring or denying them in their surrender to uniformist concepts of earth history. For example, Russell L. Mixter of Wheaton College asserted that "an honest creationist will ask the paleontologist what he knows of the time of origin of animals, and draw his conclusions from the data."[40] When the paleontologist Louis Leakey suggested a date of 1,750,000 years for a human fossil he discovered, Buswell predictably found no particular problem of accommodating Genesis 5 and 11 to this new chronology (again appealing to Warfield). In "correspondence with other creationist anthropologists in

40. *Evolution and Christian Thought Today*, p. 183.

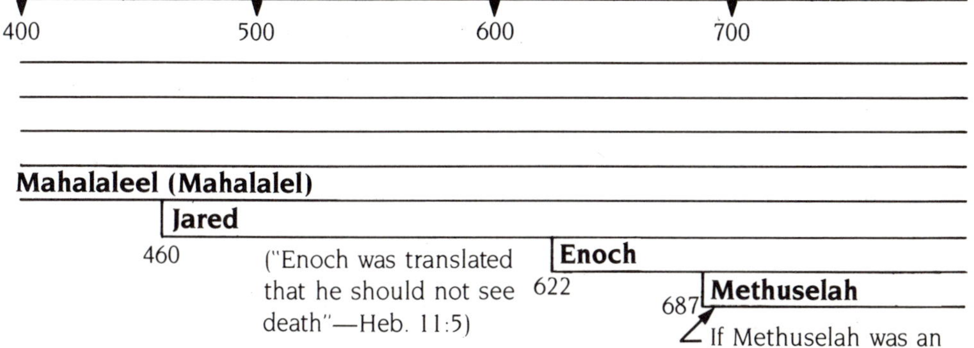

If Methuselah was an actual son rather than a mere descendant of godly Enoch, the probability would be greater that he was also a righteous man. But if Noah was the only righteous man on earth *before* the Flood year began (Gen. 6:9, 18; 7:1), Methuselah could not, as a righteous (i.e., justified) man, have lived to the very year of the Flood as the no-gap interpretation of Genesis 5 and 11 would require. Thus, Methuselah would be an ancestor, not the actual father, of Lamech; and more than 1,656 years would have elapsed between Creation and the Flood, thus providing an analogy to Genesis 11 where it is even more clear that more than the minimum of 292 years elapsed between the Flood and the birth of Abraham.

direct solicitation for opinion about *Homo habilis* . . . the responses indicated a general lack of alarm at the increased antiquity."[41]

Such men may see no problem in allowing 100,000 years between *each* of the twenty patriarchs of Genesis 5 and 11, but for most Bible-believing Christians this is an utter absurdity. Even as our understanding of *the dignity and supernatural creation of man* rests upon the clear terms of special revelation, so also our understanding of *the basic outline of man's earliest history* must come from Scripture rather than from science.[42]

The fact that the first eleven chapters of Genesis cannot be harmo-

41. *Journal of the American Scientific Affiliation* 17:3 (September 1965), p. 77.

42. For helpful discussions of human fossils, see Arthur C. Custance, "Fossil Man in the Light of the Record in Genesis," in *Why Not Creation?* ed., Walter Lammerts (Nutley, N.J.: Presbyterian and Reformed Pub. Co., 1970), pp. 194–229; Ian T. Taylor, *In The Minds of Men*, pp. 204–64; Michael Pitman, *Adam and Evolution*, pp. 86–100; Wayne Frair and Percival Davis, *A Case For Creation*, 3rd ed. (Chicago: Moody Press, 1983), pp. 117–26.

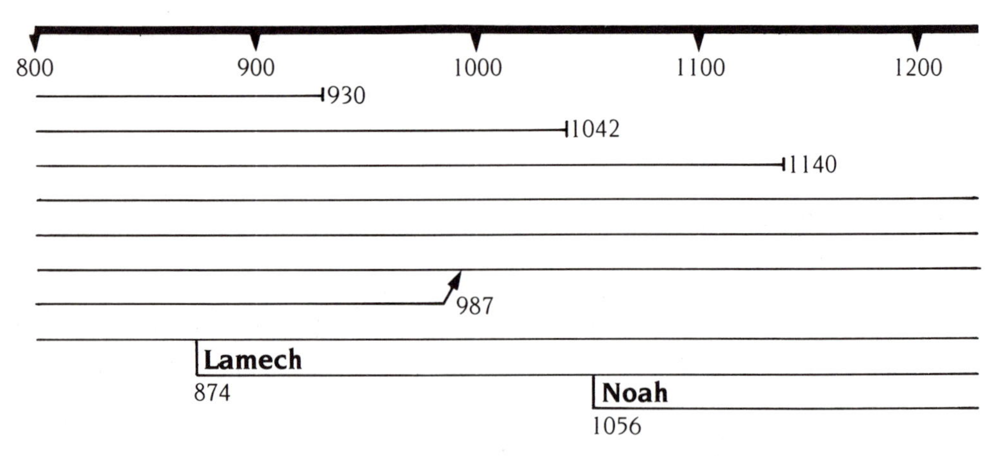

The great ages of men before the Flood cannot be modified by assuming, for example, that "years" before the Flood were only one tenth as long as our years. This would indeed make Methuselah a "reasonable" 97 years old at death. But it would also make Mahalalel and Enoch fathers at the age of six! Obviously, the Scriptures expect these large numbers to be taken at face value.

The famous Sumerian King List (c. 2000 B.C.) lists eight kings each of whom is said to have ruled for an average of 30,000 years before the Flood. "[Then] the Flood swept over [the earth]." After the Flood, the reigns of kings are listed as being much lower. This must be a highly exaggerated oral tradition of the great longevity of pre-Flood men as correctly preserved in the inspired text of Genesis 5.

---

nized with evolutionary schemes of earth history is evidenced by the fact that neo-orthodox and neo-liberal scholars have long since given up the effort of taking those chapters as serious history.[43] It is the privilege of these men to dispense with an historical Adam if they so desire. But they do not at the same time have the privilege of claiming that Jesus Christ spoke the truth. Adam and Jesus Christ stand or fall together, for Jesus said: "For if you believed Moses, you would believe me. . . . But if you do not believe his writings, how will you believe My words?" (John 5:46–47). Our Lord also insisted that . . . until heaven and earth pass away, not the smallest letter or stroke shall pass away from the Law [*and this includes Genesis*], until all is accomplished" (Matt. 5:18). If Genesis is not historically dependable, then Jesus is not a dependable guide to all truth, and we are without a Savior.

43. e.g., Ralph H. Elliott, *The Message of Genesis* (Nashville: Broadman Press, 1961).

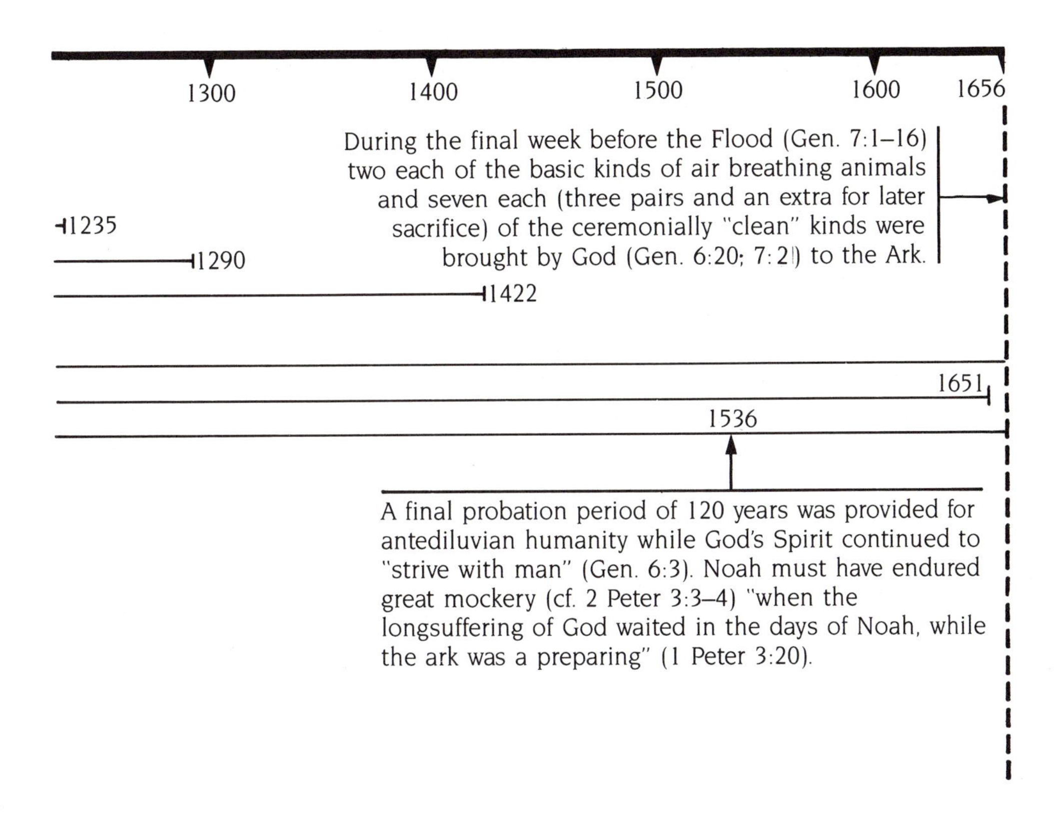

1300
1400
1500
1600
1656
During the final week before the Flood (Gen. 7:1–16) two each of the basic kinds of air breathing animals and seven each (three pairs and an extra for later sacrifice) of the ceremonially "clean" kinds were brought by God (Gen. 6:20; 7:21) to the Ark.
1235
1290
1422
1651
1536
A final probation period of 120 years was provided for antediluvian humanity while God's Spirit continued to "strive with man" (Gen. 6:3). Noah must have endured great mockery (cf. 2 Peter 3:3–4) "when the longsuffering of God waited in the days of Noah, while the ark was a preparing" (1 Peter 3:20).

The Apostle Paul said that "for as through the one man's disobedience the many were made sinners, even so through the obedience of the One the many will be made righteous" (Rom. 5:19); and "For as in Adam all die, so also in Christ all shall be made alive" (1 Cor. 15:22). If Adam didn't fall from original righteousness, then there is no sin, and Christ died for nothing. If universal death through Adam is a myth, then so is the doctrine of the resurrection, and the Apostle Paul is a false witness (1 Cor. 15:15). The full historicity of the Genesis account of Adam and Eve is absolutely crucial to the entire God-revealed plan of salvation.

Conclusion

Adam and Eve were supernaturally and suddenly created by God; they did not gradually evolve from the animal kingdom. They were created physically; not just spiritually. They were created in the image and likeness of their personal, infinite, eternal Creator; not like animals which lack that image. They were created by God to exercise total dominion as "king and queen" over a perfect and complete world; they did not rule as subordinates to a fallen angel (Satan) in a world full of the marks of destruction, death or decay. They were not only unique in their physical perfection in all of human history, but were also the most brilliant human beings the world will ever know, apart from Christ, their incarnate Creator, until the day of glorification when all of God's people will be like Him when they see Him as He really is (1 John 3:2).

Surely the words of rebuke given by our Lord to the two on the road to Emmaus must be applicable to many Christians today: "O foolish men and slow of heart to believe in all that the prophets have spoken" (Luke 24:25). Our basic problem today in the question of origins is not so much that we are ignorant of the theories and speculations of men. Our problem too often is that we know neither the Scriptures nor the power of God, and therefore deeply err in communicating God's message to modern man. May God be pleased to grant to each of us a renewing of our minds through submission to His special revelation of truth in His infallible Word, that we might prove what is that good and acceptable and perfect will of God.

Was the Earth Once a Chaos?

The Basic Issue

Conservative students of the Bible have long debated the question of whether the original creation of the heavens and the earth is to be understood as an event within the first day of creation, or whether a vast period of time could have elapsed between the original creation of Genesis 1:1 and the "formless and void" condition described in Genesis 1:2. Most Christians who favor a time gap between these two verses believe that the original earth was populated with plants and animals (and perhaps even pre-Adamic "men"), and because of the fall of Satan it was destroyed by God through means of a global flood, was plunged into total darkness, and thus *became* "formless and void." The supposedly vast ages of the geologic timetable are thought to have occurred during this interval, so that the fossil plants and animals which are found in the crust of the earth today are relics of the originally perfect world which was supposedly destroyed *before* the six literal days of creation (or, rather, re-creation) as recorded in Genesis 1:3–31.

The Gap Theory (or Ruin-reconstruction Theory) has been widely accepted among evangelical Christians, especially since the early nineteenth century when Thomas Chalmers of Scotland popularized this interpretation, presumably with the motive of harmonizing the Genesis account of creation with the vast time periods of earth history demanded by uniformitarian geologists.[1] The theory was elaborated in 1876 by George H. Pember (*Earth's Earliest Ages*), and then enormously popularized in the footnotes of the *Scofield Reference Bible* beginning in 1917. In 1970, Arthur C. Custance, a Canadian scientist, published a defense of the Gap Theory entitled, *Without Form and Void*.

The differences between the Gap Theory and the traditional view of a comparatively recent creation of the earth within six literal days are quite profound. In the *first* place, the Gap Theory must redefine the "very good" of Genesis 1:31 ("And God saw all that He had made, and behold, it was very good"), for Adam would have been placed as a very late arrival in a world that had already been destroyed, so that he was literally walking upon a graveyard of billions of creatures (including dinosaurs) over which he would never exercise dominion (Gen. 1:26). Futhermore, this "very good" world would already have become the domain of a fallen and wicked being who is described elsewhere in Scripture as "the god of this world" (2 Cor. 4:4).

Secondly, the Gap Theory assumes that carnivorous and other animals were living and dying not only before Adam, but even before the fall of Satan. But could death have prevailed in the animal kingdom in a sinless world? Does not the Bible indicate that "the creation was subjected to futility," and that it "groans and suffers the pains of childbirth" as a result of the Edenic Curse, which came *after* Adam's fall (Rom. 8:20–22)? It was neither nature nor Satan, but *man* who was created to be the king of the earth (Ps. 8, Heb. 2:5–8); and not until *man* deliberately rejected the known will of God did death make its first appearance on this planet (Rom. 5:12) or did animals fall under the "bondage of corruption" (Rom. 8:21). Thus, the Gap Theory seriously compromises the biblical doctrine of man's original dominion and the doctrine of the Edenic Curse which a holy God inflicted upon the earth because of man's rebellion.

Thirdly, if, according to the Gap Theory, all the animals and plants of the "first world" were destroyed and fossilized they could have had no

1. Although the Gap Theory had been advocated in one form or another spasmodically for centuries (see documentation in Arthur Custance, *Without Form and Void*, Box 291, Brockville, Ontario, Canada, 1970), it was first popularized by Dr. Thomas Chalmers of Edinburgh University in 1814. In this way he attempted to incorporate Georges Cuvier's concepts of geologic catastrophism into a biblical framework. See *The Works of Thomas Chalmers on Natural Theology* (Glasgow: Wm. Collins and Co., n.d.); Bernard Ramm, *The Christian View of Science and Scripture* (Grand Rapids: Eerdmans, 1954), pp. 195ff.; Francis Haber, *The Age of the World* (Baltimore: Johns Hopkins Press, 1959), pp. 201ff.; Erich Sauer, *The King of the Earth* (Grand Rapids: Eerdmans, 1962), pp. 230ff.

genetic relation to the living things of the present world, in spite of the fact that the majority of them appear to be identical in form to modern types. Likewise, those who place human fossils into this "gap" period are forced to the conclusion that such pre-Adamic "men" did not possess an eternal soul (because they obviously died before sin entered the world by Adam).[2]

Fourth, the Gap Theory leaves us with no clear word from God concerning the "original perfect world" (which most advocates of this theory assume to have existed for many millions of years). Thus, we would know *nothing* of the order of events in its creation, the arrangement of its features, or its history (which, we are told, could have constituted 99.9 percent of earth's history thus far); for instead of having the entire first chapter on this important subject, we have only the first verse! Are Christians to assume that before Genesis 1:2 we must look to uniformitarian and evolutionary geologists to fill in the blank? What does this do to Exodus 20:11 (cf. 31:17), which states that *within the six days* (not before the first day), God "made the heavens, the earth, the sea, and all that is in them" (not just plants, animals, and men)?

Finally, the Gap Theory tacitly assumes that Noah's Flood (to which Moses devotes three entire chapters in Genesis) was comparatively insignificant from the standpoint of its geologic and hydrodynamic effects, since all the major fossil-bearing formations were laid down by the supposed Flood of Genesis 1:2 (sometimes referred to as Lucifer's Flood). Obviously, the same fossils were not deposited by two universal floods separated by a long time period. Therefore, the Gap Theory almost requires that Noah's Flood be localized in its extent and effect in order to give full emphasis to a supposed pre-Adamic catastrophe (cf. Whitcomb and Morris, *The Genesis Flood*, pp. 5–6). It is futile to argue that the Apostle Peter was referring to a catastrophe back in Genesis 1:2 when he wrote that ". . . the world at that time was destroyed, being flooded with water" (2 Peter 3:6), for he had just referred to Noah's Flood (2 Peter 2:5) and would hardly be referring to a different Flood without explanation, especially since the only Flood the Lord Jesus Christ ever spoke of was in Noah's time (cf. Matt. 24:37–39; Luke 17:27).

Obviously, then, the Gap Theory is not simply a minor deviation from the traditional interpretation of the Genesis creation account. For this

2. For defenses of the pre-Adamic race view, see Gleason L. Archer, A *Survey of Old Testament Introduction* rev. ed. (Chicago: Moody Press, 1974), pp. 196–99; and Charles F. Baker, *A Dispensational Theology* (Grand Rapids: Grace Bible College Publication, 1971), p. 207.

Mountains, Ice and Snow

Mountain ranges in our present world are vastly different from those before the Flood. In the first place, they are as much as *four times* higher in elevation, some being over 28,000 feet above sea level. Such mountains could never have been covered by a global Flood; however, if the present earth, with its 300,000,000 cubic miles of water, had all its surface features flattened out, this water would cover the earth to a depth of 12,000 feet. Secondly, they are packed with billions of fossils of plants and animals that were rapidly buried to great depths by the swirling waters of the great Flood. Mountains before the Flood had *no* fossils, for they were uplifted by God before living things were created (Gen. 1:9–10, 20–22). Thirdly, they are covered with snow and ice. Before the Flood, the great vapor canopy (Gen. 1:6–8) produced a greenhouse effect, trapping the reflection of solar heat (as on Venus today) and providing a warm climate even in the polar regions. The collapse of this vapor canopy during the early weeks of the Flood (Gen. 7:11–12) took the form of snow and ice in the higher latitudes, causing huge glaciers, the sudden freezing of mammoths and other creatures, and the locking up of enough water in the form of ice to expose land bridges from Asia to Alaska and Australia. This intense, but comparatively brief, "ice-age" following the Flood has been considerably modified in recent millennia, causing ocean levels to rise and many sea-mounts and land bridges to be drowned. See Joseph C. Dillow, *The Waters Above* and J. C. Whitcomb, *The World That Perished.*

reason, the biblical evidences that have been set forth in its defense need to be carefully examined. Probably the four most significant arguments in support of the Gap Theory are these: (1) The verb translated "was" in Genesis 1:2 (Hebrew: *hāyᵉṯâ*) should better be translated "became" or "had become," thus permitting the idea of a profound change in the earth's condition. (2) The phrase "formless and void" (Hebrew: *ṯōhû wāḇōhû*) appears elsewhere only in Isaiah 34:11 and Jeremiah 4:23, and the context of those passages speak clearly of judgment and destruction. Futhermore, the word *ṯōhû* by itself frequently has an evil connotation. (3) It is highly improbable that God, the author of light, would have originally created the world in darkness, which is generally used in Scripture as a symbol of evil. (4) There seems to be a definite distinction in the first chapter of Genesis between "created" and "made," thus permitting us to assume that many of the things mentioned throughout Genesis 1 were simply recreated.

"Was" or "Became"?

The first supporting argument for the Gap Theory (and the one which Arthur Custance considers crucial) is that the Hebrew verb *hāyᵉṯâ* in Genesis 1:2 should be translated "became" or "had become," thus implying a tremendous transition from perfection to judgment and destruction.

The answer to this argument is that while the verb *hāyᵉṯâ* can often be translated "became," the word order and sentence structure in Genesis 1:2 (and in a number of other passages) does not permit this translation. If it *had* to be translated "became," then we would *have* to say that Adam and Eve "became" naked (Gen. 2:25), and that the serpent "became" more subtle than any beast of the field (Gen. 3:1)!

The word order in Genesis 1:2 (subject then verb) is most often employed to signal the addition of circumstantial information and the absence of sequential or chronological development, and that is why the Septuagint translators rendered the verb "was" and not "became." Furthermore, the Hebrew word *waw* which begins Genesis 1:2 is a "circumstantial *waw*" because it is attached to the subject ("the earth") rather than to the verb. Thus, it should properly be translated "now," and the Septuagint translators, who were extremely careful in their handling of the Pentateuch, gave it this translation (*de*).[3]

3. Charles R. Smith, in a review of Arthur C. Custance, *Without Form and Void*, the September 1971 (8:2) issue of *Creation Research Society Quarterly* (2717 Cranbrook Road, Ann

Very illuminating parallels to the construction in Genesis 1:2 are found in Zechariah 3:1–3 (". . . he showed me Joshua. . . . Now Joshua was clothed with filthy garments. . . .") and Jonah 3:3 ("So Jonah arose, and went into Nineveh. . . . Now Nineveh was an exceedingly great city. . . ."). Obviously, Joshua did not become clothed with filthy garments *after* Zechariah saw him; nor did Nineveh become a great city *after* Jonah entered it. Thus, all the important English translations of Genesis 1:2 are correct in avoiding the idea of "became," because the verse is simply describing the earth's condition just after it was created. In the light of this context and word order, the following theological monstrosity would be produced in Genesis 1:2 if one were to insist on the idea of change or transition in the verb *hāyetâ*: "Now at the time when God created, the earth had become [already, prior to its creation] unformed and uninhabited"!

"Empty" or "Chaotic"?

This brings us to the second important argument used in support of the Gap Theory. If Genesis 1:2 describes the earth's condition at the time of creation, how do we explain the phrase "formless and void" (*tōhû wābōhû*)? Would an infinitely wise and powerful God have created the earth in such a chaotic condition? The only other places in the Bible where the two words *tōhû* and *bōhû* appear together (Isa. 34:11 and Jer. 4:23) are passages that speak of divine judgment upon Gentile nations and upon Israel. Does not this indicate that these words must refer to judgment and destruction in Genesis 1:2? Even the word *tōhû* (translated "without form" in the KJV and "formless" in the NASB), in the twenty verses where it appears without *bōhû* in the Old Testament, is sometimes used in an evil sense.

This is admittedly an impressive argument, for one of the most dependable ways to determine the meaning of Hebrew words and phrases is to compare their usage in other passages. Thus, if *tōhû* refers to some-

---

Arbor, Mich. 48104). In a published debate on the Gap Theory ("And the Earth Was Without Form and Void," *Journal of the Transactions of the Victoria Institute*, Vol. 78 (1946), pp. 21–23), F. F. Bruce, pointed out that if Genesis 1:2 indicated an event subsequent to the creation of verse 1, we might have expected in verse 2 a "*waw* consecutive" with the imperfect tense (i.e., *wattehī/hā'āreṣ*, not *wehā'āreṣ/hāyetâ*. See also E. A. Speiser, *The Anchor Bible*, Vol. 1: *Genesis* (New York: Doubleday and Company, Inc., 1964), p. 5.

thing evil when used elsewhere in the Old Testament, it would probably have this connotation in Genesis 1:2. But a careful examination of the usage of this word does not support such a meaning. For example, in Job 26:7 we read that God "'stretches out the north over empty space [*t̠ōhû*], and hangs the earth on nothing.'" Certainly we are not to find in this verse any suggestion that outer space is basically evil! In some passages the word refers to the wilderness or desert, which is conspicuous for the absence of life (Deut. 32:10; Job 6:18; 12:24; Ps. 107:40). In most of the places where the word appears in Isaiah, it is paralleled with such words as *nothing* and *nought*.

Of particular interest in this connection is Isaiah 45:18, which has been used as an important proof text for the Gap Theory. The verse tells us of "the God who formed the earth and made it, He established it and did not create it a waste place [*t̠ōhû*], *but* formed it to be inhabited." It has been claimed that the *t̠ōhû* condition of the earth in Genesis 1:2 could not have been its original condition, because Isaiah 45:18 says it was *not* created a *t̠ōhû*. Consequently, God must have originally created an earth replete with living things, and later destroyed it, causing it to *become* "*t̠ōhû*."

However, such an interpretation overlooks the true significance of the final phrase in this verse: "formed it to be inhabited." The real point of the passage seems to be that God did not ultimately intend that the world should be devoid of life, but rather that it should be filled with living things. Thus, He did not allow it to *remain* in the empty and formless condition in which He first created it, but in six creative days filled it with living things and fashioned it to be a beautiful home for man. The verse thus speaks of God's *ultimate purpose* in creation, and the contrast in this verse between *t̠ōhû* and "inhabited" shows clearly that *t̠ōhû* means "empty" or "uninhabited," rather than "judged," "destroyed," or "chaotic." Arthur Custance frankly confesses that "Isaiah 45:18 is a strong witness only to those who already accept the alternative rendering of Gen. 1:2" (*Without Form and Void*, p. 115), especially because the word *t̠ōhû* appears again in the following verse (Isa. 45:19) and can hardly be translated "ruin" in that context.

To be sure, the only passages besides Genesis 1:2 where *t̠ōhû* and *bōhû* appear together—Isaiah 34:11 and Jeremiah 4:23—are placed in contexts which emphasize divine judgment. But even here the basic meaning of *empty* or *uninhabited* fits well. Since God's ultimate purpose for the earth, and particularly the Holy Land, was that it might be *filled with people* (Isa. 45:18; 49:19–20; Zech. 8:5), it would be clear evidence of His wrath and displeasure for the Promised Land to become *empty* and *uninhabited* again. The concept of emptiness, therefore, implies divine judgment only when it speaks of the removal of something that is good. On the other hand, when emptiness follows something that is evil it can be a comparative blessing. An example of this may be found in Christ's work of casting demons out of people (Luke 8:27–35; cf. Matt. 12:44—"unoccupied, swept, and put in order").

In spite of the fact that the phrase *ṯōhû wābōhû* appears elsewhere in judgment contexts and thus takes on an evil connotation in those passages, the same phrase may have a very different connotation when it appears in a different context. Even advocates of the Gap Theory admit that a context of divine judgment seems to be missing in the opening verses of Genesis.[4] It is true that the earth was *empty* as far as living things are concerned, and it was devoid of many of the magnificent, majestic, and beautiful features it later possessed, such as continents, mountains, rivers, and seas; but it was certainly not *chaotic*, *ruined*, or *judged*. Edward J. Young feels,

> It would probably be wise to abandon the term 'chaos' as a designation of the conditions set forth in verse two. The three-fold statement of circumstances in itself seems to imply order. The material of which this earth consists was at that time covered with water, and darkness was all about. Over the waters, however, brooded God's Spirit.[5]

So far from being chaotic, the earth *at this particular stage* of creation week can be described as *perfect*. There was nothing wrong with any of the material elements that God brought into existence. The earth had a core, mantle, and crust composed of perfect metal and rock; it was covered with oceans of perfect water; and it was surrounded by a blanket of perfect atmosphere. But it was not yet *complete* as far as God's ultimate purposes were concerned. Likewise, Adam, as a man, was *perfect* when he was first created. But he was "alone" and to this extent *incomplete* until God

4. J. H. Kurtz, *Manual of Sacred History*, 1888, p. xxvi. Cited by Curtis C. Mitchell, "A Biblical and Theological Study of the Gap Theory" (unpublished Th.M. thesis for Talbot Theological Seminary, La Mirada, Calif., 1962), p. 45.

5. Edward J. Young, *Studies in Genesis One*, p. 13. Thus, we have an important alternative to the only two interpretations of Genesis 1:2 suggested in the *New Scofield Reference Bible* (p. 1, note #5). In addition to the "Original Chaos" and "Divine Judgment" interpretations suggested there, we have what must be considered the traditional Jewish and Christian interpretation, namely, the "Originally Perfect Yet Incomplete" view.

created Eve to be his companion. For this reason, God could describe Adam's pre-Eve condition as "not good" (Gen. 2:18). In other words, until the creation week ended, Adam himself was *ṯōhû wābōhû* (perfect at this stage of creation, but alone, incomplete, and thus *comparatively speaking* "not good").

Was the Darkness Evil?

The third major argument used in support of the Gap Theory concerns the darkness of Genesis 1:2. Since darkness is almost always used as a symbol of sin and judgment in the Scriptures (John 3:19; Jude 13, etc.), and since God did not say that the darkness was "good" (as He did concerning the light—Gen. 1:4), proponents of the Gap Theory insist that God originally created the world in light (Ps. 104:2; 1 Tim. 6:16) and only later plunged it into darkness because of the sin of Satan and other angels.

This, again, is an impressive argument. But *all* of the biblical evidences need to be taken into consideration. Psalm 104:19–24, for example, makes it quite clear that *physical* darkness (absence of visible light) is not to be considered as inherently evil or as the effect of divine judgment. Speaking of the wonders of the day-night cycle, the Psalmist states:

> . . . The sun knows the place of its setting. Thou dost appoint darkness and it becomes night, in which all the beasts of the forest prowl about. The young lions roar after their prey, and seek their food from God. . . . O LORD, how many are Thy works! In wisdom Thou hast made them all; the earth is full of Thy possessions.

If the appointing of physical darkness is a revelation of God's wisdom and riches, how can it be inherently evil?

In discussing the opening verses of Genesis, Edward J. Young pointed out the true significance of the term *darkness*:

> God gives a name to the darkness, just as he does to the light. Both are therefore good and well-pleasing to him; both are created, although the express creation of the darkness, as of other objects in verse two, is not stated, and both serve his purpose of forming the day. . . . Darkness is recognized in this chapter as a positive good for man. . . . Whatever be the precise connotation of the [evening] of each day, it certainly included darkness, and that darkness was for man's good. At times, therefore,

darkness may typify evil and death; at other times it is to be looked upon as a positive blessing.[6]

The reason why God did not "see that the darkness was good" may be that darkness is not a specific entity, or a thing, but it is rather an absence of something, namely, light. Perhaps it is for this same reason that God did not see that the "expanse" of the second creative day was good. It, too, was a negative entity, being the empty space between the upper and lower waters. The fact that the absence of light is not incompatible with the presence and blessing of God is evidenced by the statement that "the Spirit of God was moving over the surface of the waters" in the midst of this primeval darkness. In the words of the Psalmist, "Even the darkness is not dark to Thee, and the night is as bright as the day. Darkness and light are alike to Thee" (Ps. 139:12).

How Many Creative Acts in Genesis 1?

The fourth major supporting argument for the Gap Theory is built upon a supposed distinction between the verbs "created" (*bārā'*) and "made" (*'āśâh*). If this distinction is not clearly maintained, then the Gap Theory must collapse, for Exodus 20:11 states, "For in six days the LORD made the heavens and the earth, the sea and all that is in them." Obviously, if God "made" *everything* within six days, there would be no room for a long time interval between the creating of the heavens and earth (Gen. 1:1) and the creating of all the other things (Gen. 1:2–31). Therefore the Gap Theory requires that "made" (*'āśâh*) in Exodus 20:11 should be understood as referring only to the "refashioning" of the heavens and earth in six days after the supposed judgment of Genesis 1:2.

It is a disappointment to find the outstanding *New Scofield Reference Bible* (1967) supporting this distinction between "created" and "made" in Genesis 1—"Only three creative acts of God are recorded in this chapter: (1) the heavens and the earth, v. 1; (2) animal life, vv. 20–21; and (3) human life, vv. 26–27. The first creative act refers to the dateless past" (p. 1, note #4). With regard to Genesis 1:3 ("Then God said, 'Let there be light;' and there was light"), the *New Scofield Reference Bible* states that "neither here nor in vv. 14–18 is an original creative act implied. A different word is

6. Ibid., pp. 21, 35.

used. The sense is *made to appear, made visible*. The sun and moon were created 'in the beginning.' The light came from the sun, of course, but the vapor diffused the light. Later the sun appeared in an unclouded sky" (p. 1, note #6).

But this interpretation raises serious questions. In the first place, if God had intended to convey to us the idea that the heavenly bodies (sun, moon, and stars) were already in existence on the first day, but only "appeared" on the fourth day (by a removal of clouds) the verb *to appear* could easily have been used, as in Genesis 1:9 ("and let the dry land appear"). Furthermore, if the creation of the sun occurred as part of the creative activity supposedly covered by Genesis 1:1, how could the earth have been shrouded in total darkness in 1:2? No cloud canopy could have excluded the sun's light, for water vapors were not elevated above the firmament until the *second* day of creation.

Even more serious for the Gap Theory is the fact that Genesis 1:21 states, "And God created [*bārā'*] the great sea monsters. . . ." while verse 25 states, "And God made [*'āśâh*] the beasts of the earth. . . ." Surely we are not to think that sea creatures were directly "created" on the fifth day, but land animals were merely "appointed" or "made to appear" on the sixth day! All those who hold that *bārā'* and *'āśâh* cannot be used of the same kind of divine activity are faced with a serious difficulty here. In fact, the difficulty is so severe that the *New Scofield Reference Bible*, in support of this distinction, suggests that the beasts which were "made" on the sixth day (vs. 25) were actually already "created" on the fifth day (p. 2, note #2). But such an interpretation is impossible since the beasts were obviously brought into existence for the first time on the sixth day ("let the earth *bring forth*," vs. 24). This bringing into existence is described as a work wherein God "*made* the beasts of the earth" (vs. 25).

And what does the Gap Theory do about the plant kingdom, which was "brought forth" from the earth on the third day (vss. 11–12)? It must reject the idea that it was created on that day! "It is by no means necessary to suppose that the life-germ of seeds perished in the catastrophic judgment [Gen. 1:2] which overthrew the primitive order. With the restoration of dry land and light the earth would 'bring forth' as described" ("Old" *Scofield Reference Bible*, p. 4, note #3). But this is a bizarre concept, especially when we realize how rich is God's use of synonyms for "created" in these passages. For example, He commanded the waters to "teem with swarms of living creatures" (vs. 20). This is explained in the following verse to mean that "God created [*bārā'*] . . . every living creature that moves, with which the waters swarmed." Likewise, Genesis 2:7 tells us that "God formed [*yāṣar*] man of dust from the ground," which must mean "created," in the light of Genesis 1:27.

Arthur Custance, in his book, *Without Form and Void*, even attempts to draw a distinction between God's *making* us in His image and likeness (1:26) and *creating* us in His image (1:27). Appealing to Origen (third century A.D.), he concludes that "while both image and likeness were *appointed* (*'āśâh*), only the image itself was *created* (*bārā'*) by God, the achievement of the likeness being left as something to be wrought out by experience" (p. 180). Thus, according to Custance, we were not created in the image *and* likeness of God! But if this be true, then God could hardly have *made* man in His *likeness* in the same day He *created* him (Gen. 5:1). Also, one wonders how Adam could have begotten Seth "in his own likeness" as well as "according to his image" (5:3).

These examples should suffice to show the absurdities to which one is driven by making distinctions which God never intended to make. For the sake of variety and fullness of expression (a basic and extremely helpful characteristic of Hebrew literature), different verbs are used to convey the concept of supernatural creation. It is particularly clear that whatever shade of meaning the rather flexible verb *made* (*'āśâh*) may bear in other contexts of the Old Testament, in the context of *Genesis* 1 it stands as a synonym for *created* (*bārā'*).[7] Thus, not only animal life and human life, but also plant life and the astronomic bodies were directly created by God in their appropriate days; and this fact, in the light of Exodus 20:11, is utterly devastating to the Gap Theory.

Other Gap Theory Arguments

In addition to the four major arguments for the Gap Theory discussed above, one frequently hears the claim that the phrase "replenish the earth" in the King James Version of Genesis 1:28 implies that the earth was once filled but now had to be filled *again* (replenished, or refilled). But the verb in the Hebrew text (*mil'û*) simply means "fill," with no suggestion of a repetition. This has been acknowledged by Custance (*Without Form and Void*, p. 8).

Some writers claim that Hebrews 4:3 should be translated "The works

7. See further discussion above, Chapter 3, Note 1.

were finished from the *downfall* of the world," linking this with the catastrophic interpretation of Genesis 1:2. But this cannot be supported by context or usage of the word (cf. Heb. 9:26).[8]

It is also frequently claimed that Ezekiel 28:13–14 demands an originally glorious world before the "formless and void" situation in Genesis 1:2, for it speaks of Satan as dwelling in ". . . Eden, the garden of God. . . . the holy mountain of God" and walking "in the midst of the stones of fire" before his rebellion against God. But it seems clear from a comparison with Daniel 2:45 and Isaiah 14:13 that "the holy mountain of God" must refer to the third heaven of God's immediate presence and not to an earthly domain. It should be noted that Satan was "cast . . . from the mountain of God . . . *to the ground*" (Ezek. 28:16–17; cf. Isa. 14:12). Apparently the Lord Jesus Christ spoke of this event when He said: "I was watching Satan fall from heaven like lightning" (Luke 10:18). It should also be noted that "Eden, the garden of God" was not a garden with trees, flowers, and streams. It was composed of precious stones and "stones of fire" (Ezek. 28:13, 14, 16). When we compare this with the description of the Holy City of Revelation 21:10–21, with its various precious stones, we conclude that Ezekiel's "garden of God" refers not to an earthly Eden back in Genesis 1:1, but to a heavenly one, from which Satan was cast down to the earth. When God created the "heavens" at the beginning of the first day of creation week, He apparently created all the angelic beings (including the unfallen Satan), who were thus on hand to sing together and shout for joy at the creation of the earth (Job 38:7). Sometime after creation week and before the temptation of Eve, Satan rebelled against his Creator. The visible earthly effect of his fall would thus not have been a catastrophe in Genesis 1:2, but the Edenic Curse of Genesis 3, which God inflicted upon the entire earth because Adam and Eve, to whom God had given full dominion of the earth, chose to believe and obey Satan rather than God (Rom. 8:20–23).

The Chaos/Creation Theory

Another variety of the Gap Theory which has gained a measure of popularity in recent years may be described as the Chaos/Creation Theory. This view posits an *originally chaotic universe* as far as the Genesis record is concerned. Genesis 1:1 is viewed as "a relative beginning rather than

8. Cf. F. F. Bruce, *The Epistle to the Hebrews* (Grand Rapids: Wm. B. Eerdmans Pub. Co., 1964), p. 71.

the absolute beginning": "The chapter would then be the accounting for the creation of the universe as *man* knows it, not *the* beginning of everything." Thus, we are asked to believe that Genesis 1

> . . . records how He brought the cosmos out of chaos . . . transformed cursing into blessing, and moved from what was evil and darkness to what was holy. . . . The clauses in verse 2 are apparently circumstantial to verse 3, telling the world's condition *when* God began to renovate it. It was a chaos of wasteness, emptiness, and darkness. Such conditions would not result from God's creative work (bārā'); rather, in the Bible they are symptomatic of sin and are coordinate with judgment.[9]

The Chaos/Creation Theory was apparently first popularized by Merrill F. Unger[10] and further elaborated by Bruce K. Waltke[11]. These scholars have rejected the traditional Gap Theory for its lack of exegetical sophistication; but they have left us with some unresolved theological problems: (1) Genesis 1 is stripped of any record of the original creation of the world; (2) the physical "darkness" of Genesis 1:2 which God made (Ps. 104:20) and named (Gen. 1:5) is transformed into something evil; (3) an alien concept called "chaos" is introduced into the creation narrative, with appeals to supposed parallels in Babylonian mythology;[12] and (4) the strong requirements of Exodus 20:11 (in the light of Col. 1:16) for the creation of all things within the six-day creation week are essentially ignored.

Merrill Unger, Bruce Waltke, and Allen Ross would surely concede an *original* creation of the universe *ex nihilo* (out of nothing). But their removal of this all-important cosmic event from Genesis 1, and the introduction of a "chaos/creation" concept (essentially unknown in Hebrew-Christian tradition) raises serious questions concerning the *when* and *how* of the original creation event. One can only hope that these Old Testament scholars would strongly endorse Carl F. H. Henry's conclusion:

9. Allen P. Ross, "Genesis," in John F. Walvoord and Roy B. Zuck, eds. *The Bible Knowledge Commentary: Old Testament* (Wheaton, IL: Victor Books, 1985), p. 28. For grammatical evidence that Genesis 1:1 (rather than 1:3) records the first creative act of God, and is an absolute (not a dependent) clause, see John J. Davis, *Paradise to Prison* (Grand Rapids: Baker Book House, 1975), pp. 39–40.

10. Merrill F. Unger, "Rethinking the Genesis Account of Creation," *Bibliotheca Sacra* 115:457 (January, 1958), pp. 27–35.

11. Bruce K. Waltke, "The Creation Account in Genesis 1:1–3," *Bibliotheca Sacra* 132:525–28 (January-October, 1975).

12. Ibid., p. 329.

Dinosaurs

The dinosaurs ("terrible lizards") flourished especially during the period from Adam to the Flood because of the warm and humid climate that characterized the entire pre-Flood world. The skeletal remains of one dinosaur ("Supersaurus") which has been discovered in Colorado suggests that he may have weighed 200,000 pounds! (See *National Geographic Magazine,* August, 1978, p. 176). They did not become extinct *before* Adam, for he was given dominion over *all* the kinds of animals (Gen. 1:28). In the broader sense of the term *dinosaur,* we may say that they are not yet extinct. On the island of Komodo in Indonesia, about a thousand huge dragon lizards still survive, some of them attaining a length of ten feet and a weight of over three hundred pounds (*National Geographic Magazine,* Dec., 1968). And surely the twenty-foot crocodile would qualify as a "terrible lizard"! Since reptiles attain sexual maturity long before their full growth is reached, we need not assume that huge and therefore old individuals represented their kind on Noah's ark. After the Flood, reptilian dinosaurs found themselves confined to a comparatively narrow belt near the equator, and thus in most cases became extinct during the subsequent centuries of struggle for existence against the more versatile and adaptable mammals. See John C. Whitcomb, *Dinosaurs and Men* (Winona Lake, IN: Grace Theological Seminary) an album of three cassettes with full outline.

> . . . Genesis 1:1 traces everything to Elohim's creative act. . . . Judeo-Christian theism . . . affirms divine creation from nothing. . . . This emphasis rules out any metaphysical principle or entity of equal ultimacy with God—be it chaos, darkness, matter, or anything else.[13]

Conclusion

The Gap Theory in its various forms continues to claim some support in the evangelical Christian community because it offers rather impressive biblical support for a position that does not radically challenge the geologic timetable of modern historical geology. Nevertheless, this theory, on closer inspection, compromises the unity and completeness of the creation account, the original perfection of the world, the genetic continuity of fossil and living forms, the totality of Adam's dominion, and the uniqueness of both the Edenic Curse and the global catastrophism of Noah's Flood.

I would agree with advocates of the Gap Theory that "the earth has undergone a cataclysmic change as a result of divine judgment. The face of the earth bears everywhere the marks of such a catastrophe" ("Old" *Scofield Reference Bible*, p. 3, note #3). However, this catastrophe must be identified with the universal Flood of Noah, which not only occupies three entire chapters of Genesis, but also is referred to by David (Ps. 29:10), Isaiah (54:9), Christ (Matt. 24:37–39; Luke 17:27), and Peter (1 Peter 3:20; 2 Peter 2:5; 3:6). It was through the vast and complex current patterns of this year-long Deluge that the living creatures of the entire world were buried and fossilized in enormous sedimentary strata that underlie every continent of the globe.[14] It is *this* catastrophe that provides for us the God-given answer to the false uniformitarianism of these last days (2 Peter 3:4) and thus effectively foreshadows the final destruction of all things by fire at the climax of the Day of the Lord (2 Peter 3:7–13).

13. Carl F.H. Henry, *God, Revelation and Authority* 6:122, 124. I wish to express my gratitude to Frederic R. Howe, Associate Professor of Systematic Theology, Dallas Theological Seminary, for his assistance in preparing this analysis of the Chaos/Creation Theory.

14. For the hydrodynamic and geologic implications of the biblical doctrine of the Flood, see John C. Whitcomb, and Henry M. Morris, *The Genesis Flood*, and John C. Whitcomb, *The World That Perished*.

Summary and Conclusion

In total harmony with the methods He employed in bringing highly complex "end products" into existence suddenly during His brief public ministry on the earth, the Son of God created the earth as a dynamic, functioning, fully equipped home for man in a very short period of time. The earth did not evolve from a "chaos" of gas and dust. Neither did it cool down from a molten mass of rock and metal. It was created by purely supernatural means during six literal days and completely furnished with all the basic kinds of living things that have ever existed, including man.

Naturalistic alternatives to the God-revealed account of origins have become increasingly untenable scientifically in recent decades as our supply of knowledge in the earth and life sciences has reached staggering proportions. *Astronomers*, with all their amazing discoveries, have still failed to explain how the earth, the sun, the moon, and the stars could have evolved into their present form by natural processes. *Geologists* and *paleontologists* have failed to explain how the huge fossil strata were laid down, how mountains arose from the sea, why dinosaurs became extinct, why all the "missing links" have disappeared, how the ice age began, and how enormous lava flows were poured out upon the earth. *Biologists* and *geneticists* have failed to explain how life could have arisen spontaneously, how the DNA code was formed, why all living creatures reproduce after their kinds, and how the evolutionary hypothesis can survive the deadly formulas of the second law of thermodynamics. *An-*

thropologists have conspicuously failed to bridge the yawning bio-cultural chasm that separates the lowest man from the highest animal.

The Bible makes it clear that the early earth was an uncontaminated, highly ordered, and completely harmonious environment for the first human beings. Nature was in harmony with man because man was in harmony with his God. At first, no animals were carnivorous. It is true that plants and fruits were eaten and thus destroyed; and bacteria caused dead vegetable matter to decay; but no animal or human blood was shed through mutual destruction and no natural catastrophes imperiled living things. Although the second law of thermodynamics (entropy) was in operation, its harmful effects upon man's world were overbalanced by God's gracious control of all physical and biological systems. Similarly, overpopulation was no threat to the biosphere because God was over, not under, His own laws (and this is still true today!). Reproductive rates in all living things were under His direction.

All of this is, of course, impossible for the mind of man to grasp today through empirical science alone, apart from special divine revelation. We are so immersed in a world that "groans and suffers" (Rom. 8:22) because of the Edenic Curse (Gen. 3:16–19), that we cannot imagine what the originally perfect earth was like apart from God's explanations in the pages of His written Word. Trapped in the pincers of the first and second laws of thermodynamics, and boxed within an apparently eternal uniformitarian system, we cannot really picture genuine creation events, or a sudden reprogramming of living things to "the bondage of corruption," or a supernaturally induced destruction and mass burial of things on a global scale.

But this is basically *our* problem, not God's. He has not only provided for us fascinating and adequate hints of these primeval realities in nature itself, but has also given us a clear and self-authenticating account of His great acts of creation and judgment in the Holy Bible. Many there are, to be sure, who would insist that a comparatively recent earth would be a deception on the part of God. Such accusations are both blasphemous and unfair. If God has *told* us of His creative methods, the order of events in the creation of various entities, and the amount of time which elapsed between these creative acts, we have no one to blame but ourselves for our ignorance. Furthermore, if men were really fair-minded in such matters, they would look at the other side of the ledger of empirical facts which not only permits but requires a recent origin of physical and biological systems.

Christians are deeply committed to the proposition that in every problem area, whether it be ultimate origins, ultimate destiny, or ultimate meanings and values and priorities, the God who has revealed Himself supremely in the Lord Jesus Christ and His written Word, cannot lie and cannot finally disappoint those who put their full trust in Him. The natural order *does* demand a Creator; and the wonder of all wonders is that this great Creator also came to earth to pay the full price of human sin and to make it possible for those who believe in Him to experience the fulness of His eternal purpose in creation. With all of its perfections, the Early Earth was therefore only a foretaste of the New Earth which God will some day create (Rev. 21:1).

Selected Bibliography

Aw, S. E., *Chemical Evolution: An Examination of Current Ideas*. San Diego: Master Books, 1982.

Andrews, E. H. *Christ and the Cosmos*. Welwyn, England: Evangelical Press, 1986.

Barker, David G. "The Waters of the Earth: An Exegetical Study of Psalm 104:1–9." *Grace Theological Journal* 7:1 (Spring, 1986), pp. 57–80.

Berkouwer, G. C. *The Providence of God*. Grand Rapids: William B. Eerdmans Publishing Company, 1952.

Blocher, Henri. *In the Beginning: The Opening Chapters of Genesis*. Downers Grove, IL: InterVarsity Press, 1984.

Bradley, Walter L. and Olsen, Roger. "The Trustworthiness of Scripture in Areas Relating to Natural Science." In *Hermeneutics, Inerrancy, and the Bible: Papers from ICBI Summit II*, edited by Earl D. Radmacher and Robert D. Preus. Grand Rapids: Zondervan, 1984.

Brand, Paul, and Yancey, Philip. *Fearfully and Wonderfully Made*. Grand Rapids: Zondervan Publishing House, 1980.

Bowden, M. *Ape-Men: Fact or Fallacy*. 2nd ed. Bromley, England: Sovereign Publications, 1977.

Carnell, Edward John. "Beware of the 'New Deism'." *His Magazine* 12:3 (December, 1951), pp. 14ff.

Chittick, Donald E. *The Controversy: Roots of the Creation-Evolution Conflict*. Portland, OR: Multnomah Press, 1984.

Clark, R. T., and Bales, J. D. *Why Scientists Accept Evolution*. Phillipsburg, NJ: Presbyterian and Reformed Publishing Company, 1966.

Clark, Robert E. D. *Darwin: Before and After*. Chicago: Moody Press, 1967.

Custance Arthur C. *Without Form and Void: A Study of the Meaning of Genesis 1:2*. Brockville, Ontario, Canada: By the author, 1970.

________. *The Mysterious Matters of Mind*. Grand Rapids: Zondervan Publishing House, 1980.

Davidheiser, Bolton. *Evolution and Christian Faith*. Phillipsburg, NJ: Presbyterian and Reformed Publishing Company, 1969.

Denton, Michael. *Evolution: A Theory in Crisis*. Bethesda, MD: Adler and Adler, Publishers, Inc., 1986.

DeYoung, Donald B. "Design in Nature: The Anthropic Principle." Impact #149 in *Acts and Facts* 14:11 (November, 1985).

________, and Whitcomb, John C. "The Origin of the Universe." *Grace Theological Journal* 1:2 (Fall, 1980), pp. 149–61.

Dillow, Joseph C. *The Waters Above: Earth's Pre-Flood Vapor Canopy*. Revised edition. Chicago: Moody Press, 1982.

Dixon, Malcolm, and Webb, Edwin. *Enzymes*. 2nd ed. New York: Academic Press, 1964.

Duyvené de Wit, J. J. *A New Critique of the Transformist Principle in Evolutionary Biology*. Kampen, Netherlands: Kok, 1965.

Fields, Weston W. *Unformed and Unfilled: A Critique of the Gap Theory*. Phillipsburg, NJ: Presbyterian and Reformed Publishing Company, 1976.

Frair, Wayne, and Davis, Percival. *A Case for Creation*. 3rd ed. Chicago, Moody Press, 1983.

Gish, Duane T. *Evolution: The Challenge of the Fossil Record*. El Cajon, CA: Creation-Life Publishers, 1985.

Grassé, Pierre-P. *Evolution of Living Organisms*. New York: Academic Press, Inc., 1977.

Grudem, Wayne A. "Scripture's Self-Attestation and the Problem of Formulating a Doctrine of Scripture." In *Scripture and Truth*, edited by D. A. Carson and John D. Woodbridge. Grand Rapids: Zondervan Publishing House, 1983. Pages 19–59.

Harris, R. Laird. "The Bible and Cosmology." *Bulletin of the Evangelical Theological Society* 5:1 (March, 1962), pp. 11–17.

Henry, Carl F. H. *God, Revelation and Authority*, Vol 6. Waco, TX: Word Books, Publisher, 1983.

Hitching, Francis. *The Neck of the Giraffe: Where Darwin Went Wrong*. New Haven, CT: Ticknor and Fields, 1982.

Hooykaas, R. *The Principle of Uniformity*. Leiden: E. J. Brill, 1963.

Hummel, Charles E. *The Galileo Connection: Resolving Conflicts Between Science and the Bible*. Downers Grove, IL: InterVarsity Press, 1986.

Kerkut, G. A. *Implications of Evolution*. New York: Pergamon Press, 1960.

Kline, Meredith G. "Because It Had Not Rained." *Westminster Theological Journal* 20:2 (May, 1958), pp. 46–47.

Klotz, John W. *Genes, Genesis, and Evolution*. St. Louis: Concordia Publishing House, 1955.

________. *Modern Science in the Christian Life*. St. Louis: Concordia Publishing House, 1961.

________. *Studies in Creation*. St. Louis: Concordia Publishing House, 1985.

Kofahl, Robert E., and Segraves, Kelly L. *The Creation Explanation*. Wheaton, IL: Harold Shaw Publishers, 1975.

Lammerts, Walter E., ed. *Why Not Creation?* Phillipsburg, NJ: Presbyterian and Reformed Publishing Company, 1970.

Layzer, David. "Cosmogony." *McGraw-Hill Encyclopedia of Science and Technology*, Vol. 3. New York: McGraw-Hill, 1960.

Leith, Thomas H. "Some Logical Problems with the Thesis of Apparent Age." *Journal of the American Scientific Affiliation* 17:4 (December, 1965), 118–22.

Lester, Lane P., and Bohlin, Raymond G., *The Natural Limits to Biological Change*. Grand Rapids: Zondervan Publishing House, 1984.

Leupold, H. C. *Exposition of Genesis*. Columbus, OH: The Wartburg Press, 1942.

Lubenow, Marvin L. *"From Fish to Gish": Morris and Gish Confront the Evolutionary Establishment*. San Diego: CLP Publishers, 1983.

Mixter, Russell, ed. *Evolution and Christian Thought Today*. Grand Rapids: William B. Eerdmans Publishing Company, 1959.

Moore, John N. *How To Teach Origins (Without ACLU Interference)*. Milford, MI: Mott Media Publishers, Inc., 1983.

Moorehead, Paul S., ed. *Mathematical Challenges to the Neo-Darwinian Interpretation of Evolution*. Philadelphia: The Wistar Institute Press, 1967.

Mora, Peter T. "The Folly of Probability." *The Origins of Prebiological Systems*. Edited by S. W. Fox. New York: Academic Press, 1965.

Morris, Henry M. *Biblical Cosmology and Modern Science*. Phillipsburg, NJ: Craig Press, 1970.

———. *History of Modern Creationism*. San Diego: Master Book Publishers, 1984.

———. *Creation and the Modern Christian*. San Diego: Master Book Publishers, 1985.

———, ed. *Scientific Creationism*. 2nd ed. El Cajon, CA: Master Books, 1985.

———, and Gary E. Parker. *What Is Creation Science?* San Diego: Creation-Life Publishers, 1982.

National Geographic Society. *The Incredible Machine*. Washington, D.C.: National Geographic Society, 1986.

Newman, Robert C., and Eckelmann, Herman J., Jr. *Genesis One and the Origin of the Earth*. Downers Grove, IL: InterVarsity Press, 1977.

Payne, J. Barton. "The Concept of 'Kinds' in Scripture." *Journal of the American Scientific Affiliation*, 10:2 (June, 1958), pp. 17–19.

Pitman, Michael. *Adam and Evolution*. London: Rider & Company, 1984.

Pun, Pattle P.-T. *Evolution: Nature and Scripture in Conflict?* Grand Rapids: Zondervan Publishing House, 1982.

Ramm, Bernard. *The Christian View of Science and Scripture*. Grand Rapids: William B. Eerdmans Publishing Company, 1954.

Ross, Allen P. "Genesis." In *The Bible Knowledge Commentary: Old Testament*. Edited by John F. Walvoord and Roy B. Zook. Wheaton, IL: Victor Books, 1985.

Steidl, Paul M. *The Earth, The Stars, and the Bible*. Phillipsburg, NJ: Presbyterian and Reformed Publishing Company, 1979.

———. "Planets, Comets, and Asteroids." In *Design and Origins in Astronomy*.

Edited by George Mulfinger, Jr. Norcross, GA: Creation Research Society Books, 1983.

Taylor, Ian T. *In The Minds of Men: Darwin and the New World Order.* Toronto: TFE Publishing, 1984.

Thaxton, Charles B., Bradley, Walter L., and Olsen, Roger L. *The Mystery of Life's Origin: Reassessing Current Theories.* New York: Philosophical Library, Inc., 1984.

Van Dyke, Fred. "Theological Problems of Theistic Evolution." *Journal of the American Scientific Affiliation* 38:1 (March, 1986), pp. 11–18.

Van Till, Howard J. *The Fourth Day: What the Bible and the Heavens are telling us about the Creation.* Grand Rapids: William B. Eerdmans Publishing Company, 1986.

Waltke, Bruce K. "The Creation Account in Genesis 1:1–3." *Bibliotheca Sacra* 132:525–28 (January-October, 1975).

Warfield, Benjamin B. "On the Antiquity and the Unity of the Human Race." In *Biblical and Theological Studies.* Phillipsburg, NJ: Presbyterian and Reformed Publishing Company, 1952, pp. 138–61.

Whitcomb, John C. *The Origin of the Solar System.* Phillipsburg, NJ: Presbyterian and Reformed Publishing Company, 1964.

________. *The World That Perished.* Grand Rapids: Baker Book House, 1973.

________. "The Science of Historical Geology." *Westminster Theological Journal* 36:1 (Fall, 1973), pp. 65–77.

________. "Contemporary Apologetics and the Christian Faith." *Bibliotheca Sacra* 134:534–37 (April, 1977 to January, 1978).

________. *The Bible and Astronomy.* Winona Lake, IN: BMH Books, 1984.

________, and Morris, Henry M. *The Genesis Flood.* Phillipsburg, NJ: Presbyterian and Reformed Publishing Company, 1961.

________, and DeYoung, Donald B. *The Moon: Its Creation, Form and Significance.* Winona Lake, IN: BMH Books, 1978.

Wilder-Smith, A. E. *The Creation of Life: A Cybernetic Approach to Evolution.* Wheaton, IL: Harold Shaw Publishers, 1970.

________. *The Natural Sciences Know Nothing of Evolution.* San Diego: Master Books, 1981.

Wilson, Clifford, and McKeon, Donald. *The Language Gap.* Grand Rapids: Zondervan Publishing House, 1984.

Young, Davis A. *Creation and the Flood: An Alternative to Flood Geology and Theistic Evolution.* Grand Rapids: Baker Book House, 1977.

________. *Christianity and the Age of the Earth.* Grand Rapids: Zondervan Publishing House, 1982.

Wysong, R. L. *The Creation-Evolution Controversy.* Midland, MI: Inquiry Press, 1976.

Young, Edward J. *Studies in Genesis One.* Phillipsburg, NJ: Presbyterian and Reformed Publishing Company, 1964.

Zimmerman, Paul, ed. *Darwin, Evolution, and Creation.* St. Louis: Concordia Publishing House, 1959.

________. "Some Observations on Current Cosmological Theories." *Concordia Theological Monthly* 24:7 (July, 1953).

Subject Index

Scripture Index